GUIDE TO ENGINEERING MATERIALS PRODUCERS

Edited by
John C. Bittence

Editorial Coordination
Fran Cverna

Acquisitions
Sandra Dunigan

Center for Materials Data

Production Project Manager
Suzanne E. Frueh

Production Project Specialist
Dawn R. Levicki

Production Assistant
Amy L. Bird

ASM INTERNATIONAL®
The Materials Information Society

First printing, February 1994

This book is a collective effort involving hundreds of technical specialists. It brings together a wealth of information from world-wide sources to help scientists, engineers, and technicians solve current and long-range problems.

Comments, criticisms, and suggestions are invited, and should be forwarded to ASM International.

Library of Congress Cataloging Card Number: 93-74318
ISBN: 0-87170-486-2

ASM International®
Materials Park, OH 440730002

Printed in the United States of America

Table of Contents

INTRODUCTION — HOW TO USE THIS DIRECTORY

This is a directory of worldwide producers of primary engineering materials — metals, plastics, elastomers, and ceramics. The directory provides two listings. One, arranged alphabetically by company name, provides the complete address and phone number for major plant locations. The other listing, arranged by primary material, allows you to find the names of all known producers of that material.

With an undertaking such as this, the question arises immediately of which companies to include and which to exclude. After studying many gray areas, we established one simple criterion. To be included in this directory, a manufacturer must produce a raw material, that is, a material supplied for the purpose of reforming or reshaping by the user.

We do, therefore, include semifinishing sources such as rerollers and producers of cast, extruded, or forged shapes. Cast bars, pressed and sintered flats, and forged rings, for example, fall into this category for metals. Producers of sheet and pultruded beams fit this definition for plastics.

However, we do not, in most cases, include fabricators in this directory. Fabricators include custom foundries, molders, and job shops. While we hold fast to this rule for metals, we allow more latitude for polymer composites and ceramics where the distinction sometimes becomes clouded. Heat treaters are not included, nor are producers of polymer-based adhesives and coatings.

Also, we have not included distributors since they do not produce the material. We assume that you will be directed to appropriate distributors or service centers by your contact with the producers listed in this directory.

The products listed in this directory fall into five major materials categories:

- ferrous metals
- nonferrous metals
- plastics and elastomers
- ceramics
- fibers and fillers

Products falling into the fifth category can be made from any base material.

Within these categories we provide further subdivisions whenever practical. For your convenience, we try to be as specific as possible when describing a company's product line. In some cases, however, detailed literature was not available to us. You will find these companies in the broader categories.

In each category, no attempt was made to single out composite, clad, or coated materials. In such cases, the material is listed under the category that applies for the base or matrix material. Producers of nonferrous welding rods, soldering and brazing metals, and hard-surfacing alloys are listed according to their materials' basic compositions.

Here is how the five categories are subdivided:

FERROUS METALS

Carbon steels. When information is available, we try to be more specific as follows:

- Carbon steels, cast (not parts)
- Carbon steels, mill shapes: rod, bar, seamless and welded tubing, free-machining bars, wire, flats, angles, beams, and other shapes
- Carbon steels, sheet and strip
- Carbon steels, plate, structural plate
- Carbon steels, forged (not parts)
- Carbon steels, powder

Alloy steels, including microalloyed steels as well as ultrahigh-strength steels. When information is available, we try to be more specific as follows:

- Alloy steels, cast (not parts)
- Alloy steels, mill shapes: rod, bar, seamless and welded tubing, wire, flats, other shapes
- Alloy steels, sheet and strip, including coated, high-strength low-alloy, drawing quality
- Alloy steels, plate, structural plate
- Alloy steels, forged (not parts)
- Alloy steels, powder

Stainless steels, including ferritic, martensitic, austenitic, precipitation-hardening, corrosion-resistant or heat-resistant grades. When information is available, we try to be more specific as follows:

- Stainless steels, cast (not parts)
- Stainless steels, mill shapes: rod, bar, tubing, wire, flats, and other mill forms
- Stainless steels, sheet and strip
- Stainless steels, plate
- Stainless steels, forged (not parts)
- Stainless steels, powder
- Ferroalloys, pure iron, including iron powder
- Iron-base superalloys
- Cast iron, including ductile, white, gray, malleable, alloy, heat resisting
- Tool steels
- Ferrous electronic, magnetic alloys, ferrites
- Ferrous welding, hard facing electrodes

NONFERROUS METALS

- Aluminum and alloys, powder
- Beryllium and alloys
- Bismuth and alloys
- Cadmium, chromium
- Cobalt, alloys, superalloys, powder
- Copper and alloys, including brass, bronze, german silver, nickel silver
- Indium
- Lead and alloys, including lead-antimony, lead bismuth, lead tin, lead base low-melting alloys
- Lithium and alloys
- Magnesium and alloys, powder
- Manganese
- Nickel, alloys, superalloys, powder
- Nuclear, heavy metals including uranium, strontium, plutonium, thorium
- Precious metals including gold and alloys, silver and alloys, platinum, iridium, rhodium, ruthenium, palladium, osmium
- Refractory metals including tungsten, tantalum, niobium, rhenium, molybdenum, hafnium, columbium
- Tin and alloys, tin-lead, tin-antimony, tin powder, white metal, tin-base low-melting alloys
- Titanium and alloys, powder
- Vanadium
- Zinc and alloys
- Zirconium

PLASTICS AND ELASTOMERS

- Acetals
- Acrylics
- Cellulosics, cellulose
- Epoxies, including modified epoxy, novolac
- Polyolefins, ethylene monomers, polyethylene, EVA, polypropylene
- Fluoropolymers, fluorocarbons, chloropolymers, polyvinylidene chloride
- Phenolics, modified phenolics
- Polyallomers
- Polyamides, aramids, nylons
- Polyamide-imides
- Polycarbonates
- Polyesters (thermoplastic), PET, PBT, liquid crystal polymers
- Polyesters (thermoset), including modified polyester, alloys, alkyds, melamine
- Polyketones
- Polyimides, maleimides
- Polyisoimides, acetylene/terminated
- Polyphenylenes, polyphenylene ether copolymer, PPO, PPS
- Polyurethanes, including amorphous
- Polystyrenes, styrene, SAN, ABS, SMA
- Polysulfones
- Polyvinyl chlorides, PVA
- Silicones
- Urea
- Thermoplastic elastomers, EPDM, neoprene, nitriles, buna-N, polybutylene

CERAMICS

When information is available, we try to be more specific as follows:

- Mechanical ceramics
- Electrical ceramics
- Refractory ceramics
- Engineering ceramics
- High alumina and "cermets"
- Carbides
- Nitrides
- Glass, quartz (not fibers, or fillers)
- Carbon, graphite, diamond (not fibers, fillers)
- Mica
- Oxides, not elsewhere classified

FIBERS AND FILLERS

- Fibers: glass, metal, boron, carbon, natural, synthetic polymer, other synthetic
- Fillers, additives: spheres, talc, calcium, glass, particulate additives (not chemicals)

Naturally, as political boundaries and markets change, a directory such as this is bound to become outdated within a few years. Even as we were compiling information for this project, changes were arriving routinely. We encourage readers to notify us of changes as they occur. We plan to update this volume regularly, so your cooperation will be appreciated.

Section I: Producer Listing

A & A Brazing Alloys Inc.

1795 Courtland Ct.
Addison, IL 60101
USA

phone: 708 953 9007
fax: 708 953 9015

Copper and alloys; Lead and alloys; Tin and alloys

AAR, Advanced Structures

Advanced Structure Div.
12633 Inkster Road
Livonia, MI 48150
USA

phone: 313 522 2000
fax: 313 522 2240

High alumina and "cermets"; Carbides

Acciaierie Valbruna SpA

Viale della Scienza Zona Industriale
Casella Postale 784
36100 Vicenza
Italy

phone: 39 0444 968 211
fax: 39 0444 563 372

Carbon steels; Alloy steels; Stainless steels; Tool steels

Accurate Forging Corp.

201 Pine Street
Bristol, CT 06010
USA

phone: 203 582 3169
fax: 302 583 3495

Copper and alloys

Acenor, S.A.

Planta de Larrondo
Apartado 1323
48080 Bilbao (Vizcaya)
Spain

phone: 34 94 453 1300
fax: 34 94 453 136

Alloy steels; Stainless steels; Tool steels

Acenor-Vitoria

Apartado 55
48970 Basauri (Vizcaya)
Spain

phone: 34 94 449 2800
fax: 34 94 449 1839

Carbon steels; Carbon steels, plate; Alloy steels; Tool steels

Acerinox, S.A.

C/ Santiago de Compostela, 100
28035 - Madrid
Spain

phone: 34 91 398
fax: 34 91 259 4

Stainless steels, sheet, strip

OTHER LOCATIONS

Acerinox
Pol. Ind. Zona Franca
Sector A, Calle 61 No. 12
08040 Barcelona
Spain
phone: 34 93 335 0692
fax: 34 93 336 0773

Acerinox
5 Rue Gay Lussac
95503 Gonesse
France
phone: 33 1 39 87 66 5
fax: 33 1 34 53 83 3

Acerinox
Postfach 3164
Poensgenstrasse 10
4018 Langenfeld
Germany
phone: 49 02173 27 070
fax: 49 02173 27 077

Acerinox
Casilla 51, Correo 7
Santiago
Chile
phone: 56 2 73 77 789
fax: 56 2 73 78 835

Acerol
Zona Industrial de Maia I
Sector VII, Rua E Lote 2-A
Maia, Moreira 4470
Portugal
phone: 351 2 948 6489
fax: 351 2 948 7520

Acerol
Via Paracelso 4
Centro Direzionale Colleoni
Palazzo Liocorno, ingresso 2
20041 Agrate Brianza (Milano)
Italy
phone: 39 039 63 91 31
fax: 39 039 63 91 39

Acerol
Riddervoldgst 4
0201 Oslo 2
Norway
phone: 47 2 55 13 21
fax: 47 2 44 93 00

Acerol
Weihermattstrasse 2
CH-5507 Mellingen
Switzerland
phone: 41 056 91 38 16
fax: 41 056 91 38 32

(continued)

Acerinox, S.A. (cont'd)

Acerol
Two University Plaza
Suite 307
Hackensack, NJ 07601-6201
USA

Siderinox
Av. Velez Sarsfield 1643-1645
CP 1285 Buenos Aires
Argentina
phone: 54 1 28 09 13
fax: 54 1 28 63 08

Inoxlee
1628 Ocean Centre
Canton Road
Tsuen wan
Hong Kong
phone: 852 736 92 20
fax: 852 736 11 12

Acieries de Bonpertuis

F-38140 Rives-sur-Fure
France

phone: 33 76 05 17 54
fax: 33 73 51 45 63

Stainless steels

Acme Steel Co.

13500 S. Perry Ave.
Riverdale, IL 60627-1182
USA

phone: 708 849 2500
fax: 708 849 8780

Alloy steels

Acme Tube

One Howard Avenue
Somerset, NJ 08873
USA

phone: 800 225 1059
fax: 908 560 3633

Stainless steels, mill shapes

Acos Villares Sede

Avenida do Estado 6116
01516 Sao Paulo
Brazil

phone: 55 11 278 7322
fax: 55 11 279 7015

Carbon steels; Alloy steels; Stainless steels; Tool steels

Admiral Metals

16 Park Street
Taunton, MA 02780
USA

phone: 508 824 5821

Copper and alloys

Advance Foundry Co.

P.O. Box 1411
107 Seminary Avenue
Dayton, OH 45401
USA

phone: 513 253 4148

Cast iron

Advanced Composite Materials Corp.

1525 S. Buncombe Rd.
Greer, SC 29651-9208
USA

phone: 803 877 0123
fax: 803 879 6615

Aluminum and alloys

AFORA (Aceros Afora S.A.)

Carretera de Zumarraga S/N
Apartado, 20
Azcoitia, Guipuzcoa 20720
Spain

phone: 34 943 81 02 00
fax: 34 943 81 56 69

Carbon steels; Carbon steels, mill shapes; Alloy steels; Alloy steels, mill shapes; Stainless steels; Stainless steels, mill shapes; Tool steels

Airtech International, Inc.

PO Box 6207
2542 East Del Amo Blvd.
Carson, CA 90749
USA

phone: 310 603 9683
fax: 310 603 9040

Epoxies; Polyesters, thermoset; Polyimides; Thermoplastic elastomers

AL Tech Specialty Steel Corp.

P.O. Box 152
Willowbrook Ave.
Dunkirk, NY 14048
USA

phone: 800 828 8600
fax: 716 366 0478

Alloy steels, mill shapes; Stainless steels, mill shapes; Tool steels

Albert Hoffmann GmbH

Elektro-Gusstahlwerk & Edelsta
Postfach 1280
D-52232 Eschweiler
Germany

phone: 49 0 24 03 798 0
fax: 49 0 24 03 798 266

Alloy steels; Stainless steels; Stainless steels, cast; Tool steels

Alcan Alluminio SpA

Bruno Buozzi 12
20090 Pieve Emanuele
Italy

phone: 39 290 7441
fax: 39 290 782155

Aluminum and alloys; Copper and alloys

OTHER LOCATIONS

Alcan Alluminio SpA
Via Vittorio Veneto, 106
20091 Bresso (MI)
Italy
phone: 39 2 614 541
fax: 39 2 614 0166

Alcan Alluminio SpA
Via Cassanese, 1
20096 Piotello (MI)
Italy
phone: 39 2 926 99101
fax: 39 2 921 61727

Alcan Alluminio SpA
Regione Grange
10013 Borgofranco D'Ivrea (TQ)
Italy
phone: 39 0125 752 011
fax: 39 0125 752 468

Alcan Aluminium Ltd.

Algoods Division
158 Sterling Road
Toronto, Ontario M6R 2B8
Canada

phone: 416 530 2600
fax: 416 530 4588

Aluminum and alloys

(continued)

Alcan Aluminium Ltd. (cont'd)

Alcan Aluminium Ltd.
1188 Sherbrooke Street West
Montreal, Quebec H3A 3G2
Canada

Alcan Smelters and Chemicals Ltd.
1188 Sherbrooke Street West
Montreal, Quebec H3A 3G2
Canada

Alcan Rolled Products Co.
2 Sheppard Ave. East
Suite 500
Willowdale, Ontario M2N 5Y7
Canada

Alcan Aluminum Corp.

Powders & Pigments Div.
901 Lehigh Avenue
Union, NJ 07083-0746
USA

phone: 908 851 4590
fax: 908 851 4597

Aluminum and alloys

OTHER LOCATION

Alcan Aluminum Corp.
100 Erieview Plaza
P.O. Box 6977
Cleveland, OH 44101
USA

phone: 216 523 6800

Alcan Wire Ltd.

Port Tennant
Swansea, West Glam SA1 8PS
Wales

phone: 44 0792 52251
fax: 44 0792 51140

Aluminum and alloys

ALCOA (Aluminum Company of America)

Chemicals Division
1501 Alcoa Building
425 Sixth Avenue
Pittsburgh, PA 15219
USA

phone: 412 553 4545

Aluminum and alloys; Electrical ceramics; Refractory ceramics; High alumina and "cermets"; Fillers, additives

OTHER LOCATIONS

Extrusion/Tube Division
P.O. Box 7500
Lafayette, IN 47903-7500
USA
phone: 317 448 3600

Forging Division
P.O. Box 5900
Cleveland, OH 44105
USA
phone: 216 641 3600

Foil Products Division
300 State Drive
Lebanon, PA 17042
USA
phone: 800 547 7716

Permatech, Inc.
911 East Elm Street
Graham, NC 27253
USA
phone: 919 578 0701

(continued)

ALCOA (Aluminum Company of America) (cont'd)

Rigid Packaging Division
900 South Gay Street
Knoxville, TN 37902
USA
phone: 615 594 4700

Specialty Metals Division
P.O. Box 772
New Kensington, PA 15068
USA
phone: 800 858 4638

Sheet & Plate Division
Sales Service Center
P.O. Box 8025
Bettendorf, IA 52722
USA

U.S. Metal System
1250 Riverview Tower
900 South Gay Street
Knoxville, TN 37902
USA
phone: 615 594 4280

SMD-Powders Division
Rockdale, TX 76567
USA
phone: 800 331 5370

Wheel Products Division
345 Fifteenth Street NW
Barberton, OH 44203
USA
phone: 216 848 5100

Alcoa/Messena

Wire, Rod & Bar Division
P.O. Box 150
Massena, NY 13662
USA

phone: 315 764 4394
fax: 315 764 4618

Aluminum and alloys

Algoma Steel Corp. Ltd.

PO Box 1400
Sault Ste. Marie, Ontario P6A 5P2
Canada

phone: 705 945 2302
fax: 705 945 4201

Alloy steels, sheet, strip

Allegheny Ludlum Corp.

1000 Six PPG Place
Pittsburgh, PA 15222-5479
USA

phone: 412 394 2800
fax: 412 394 2805

Stainless steels; Stainless steels, sheet, strip; Stainless steels, plate; Iron-base superalloys; Ferrous welding, hard facing electrodes; Nickel, alloys, superalloys

OTHER LOCATIONS

Allegheny Ludlum Steel
Bagdad
P.O. Box 565
Leechburg, PA 15656
USA
phone: 412 845 0600

Allegheny Ludlum Steel
Leechburg
P.O. Box 565
Leechburg, PA 15656
USA
phone: 412 845 0600

Allegheny Ludlum Steel
Brackenridge
River Road
Brackenridge, PA 15014
USA
phone:412 224 1000

Allegheny Ludlum Steel
Natrona
River Road
Brackenridge, PA 15014
USA
phone:412 224 1000

(continued)

Allegheny Ludlum Corp. (cont'd)

Allegheny Ludlum Steel
Vandergrift
130 Lincoln Avenue
Vandergrift, PA 15690
USA
phone: 412 567 2000

Allegheny Ludlum Steel
Wallingford
80 Valley Street
Wallingford, CT 06492
USA
phone: 203 269 3361

Allegheny Ludlum Steel
New Castle
P.O. Box 309
Route 38 West
New Castle, IN 47362
USA
phone: 317 529 9570

Allegheny Ludlum Steel
Lockport
695 Ohio Street
Lockport, NY 14094
USA
phone: 716 433 4411

Allegheny Ludlum Steel
Claremore
P.O. Box 948
Highway 66 and Lowry Road
Claremore, OK 74018
USA
phone: 918 341 8711

Allied Fibers

High Performance Fibers
Fibers Technical Center
PO Box 31
Petersburg, VA 23804
USA

phone: 804 520 3682
fax: 804 520 3388

Fibers

Allied Resinous Products, Inc.

Box 620
Clark St. & Whitney Road
Conneaut, OH 44030
USA

phone: 800 325 5176
fax: 216 593 2003

Polyolefins; Polystyrenes

AlliedSignal Aerospace Co.

Garrett Engine Division
111 South 34th St.
PO Box 5217
Phoenix, AZ 85010
USA

phone: 602 231 4664
fax: 602 231 7722

AlliedSignal Inc.

Engineering Plastics Division
PO Box 2332
Morristown, NJ 07962-2332
USA

phone: 800 222 0094
fax: 201 455 2551

Polyamides; Polyesters, thermoplastic

(continued)

AlliedSignal Inc. (cont'd)

OTHER LOCATIONS

AlliedSignal Europe N.V.
Amorphous Metals
Haasrode Research Park
Grauwmeer 1
B-3001 Heverlee (Leuven)
Belgium
phone: 32 16 39 1271
fax: 32 16 40 0474

AlliedSignal Inc.
Amorphous Metals
Room 1034-35
New World Office Bldg. East Wi
24 Salisbury Road, Tsimshatsu
Kowloon
Hong Kong
phone: 852 723 7288
fax: 852 369 8887

AlliedSignal Inc.
Amorphous Metals
6 Eastmans Road
Parsippany, NJ 07054
USA
phone: 201 581 7700
fax: 201 581 7717

Nippon Amorphous Metals Co., Ltd.
1 Nagasak-cho Azabu
Minato-ku
Tokyo
Japan
phone: 81 03 5570 6565
fax: 81 03 5570 6566

Alloy Cast Products, Inc.

700 Swenson Dr.
Kenilworth, NJ 07033
USA

phone: 908 245 2255
fax: 908 245 3267

Cobalt, alloys, superalloys

Alloy Engineering & Casting Co.

1700 West Washington
P.O. Box 9066
Champaign, IL 61821
USA

phone: 217 398 3200
fax: 217 398 3208

Stainless steels; Stainless steels, cast

Alloy Technology International, Inc.

169 Western Highway
West Nyack, NY 10994
USA

phone: 800 431 1854
fax: 914 358 5925

Tool steels; High alumina and "cermets"

Alpha Metals

600 Route 440
Jersey City, NJ 07304
USA

phone: 201 434 6778
fax: 201 434 7508

Copper and alloys; Lead and alloys; Tin and alloys

(continued)

Alpha Metals (cont'd)

OTHER LOCATIONS

Alpha Fry SARL
112 rue Vaillant Couturier
93130 Noisy-Le-Sec
France

Alpha Grillo Lotsysteme GmbH
Postfach 11 01 26
Buschstr. 95
4100 Duisburg 11
Germany

Alpha Grillo Lotsysteme GmbH
NLDresden
OverbeckstrBE 4
0-8030 Dresden 84-10
Germany

Alpha Metals (Korea) Ltd.
3/F Hyosuh Building
915-12, Band Bai Dong
Seochu Ku
Seoul
Korea

Alpha Metals Inc.
75 Gilkreast Rd.
Suite 200
Londonderry, NH 03053
USA

Alpha Metals Inc.
2155 Stonigton Ave.
Suite 215
Hoffman Estates, IL 60195-5209
USA

Alpha Metals Inc.
No. 4, Lane 100
Sec. 2
Nan Shan Road
Lu-Chu Hsiang, Taoyuan
Taiwan

Alpha Metals Inc.
200 Technology Drive
Alpharetta, GA 30201
USA

Alpha Metals Inc.
2751 El Presidio St.
Carson, CA 90810
USA

Alpha Metals Ireland
Unit 48
Airways Int'l Estate
Cloughran
Dublin 17
Ireland

Alpha Metals Ltd.
Block A, 1/FL
21 Tung Yuen Street
Yau Tong Bay
Kowloon
Hong Kong

Alpha Metals Ltd.
1, The Broadway
Tolworth
Surbiton, Surrey KT6 7DQ
England

(continued)

Alpha Metals (cont'd)

Alpha Metals PTE Ltd.
14 Tuas Avenue 10
Singapore 2263
Singapore

Consolidated Alloys
17-19 Trawalia Avenue
Thomastown, Victoria 3074
Australia

Alpha Metals SpA
via Trento, 65
20021 Bollate MI
Italy

Hitech Commercial e Industrual LTDA
Av. Eng. Luiz Carlos Berrini 801
04571 Brooklin Sao Paulo
Brazil

Alpha Yunnan Tin Solder Co. Ltd.
Buo Luo Cun
Chuan Jin Road
Kunnming City, Yunnan Province
People's Republic of China

Reycom Electronica S.A.
B. de Trigoyen 972 piso 2C
1304 Capital Federal
Argentina

AlSiMag Technical Ceramics, Inc.

Box 89
Highway 14
Laurens, SC 29360
USA

phone: 800 845 9761
fax: 803 682 1141

Electrical ceramics; Oxides

Aluminium Francais

4 Rue du Parc
F93155 le Blanc Mesil Cedex
France

phone: 33 1 48 67 13 57
fax: 33 1 48 67 14 67

Aluminum and alloys

Aluminium Walzwerke Singen GmbH

Postfach 160
D7700 Singen/HTWL
Germany

phone: 49 077 3180 0
fax: 49 077 3180 2222

Aluminum and alloys

Aluminum Company of Canada, Ltd.

1188 Sherbrooke St. West
Montreal, Quebec H3A 3G2
Canada

phone: 514 848 8000

Aluminum and alloys

Aluminum Smelting & Refining Co., Inc.

Certified Alloys Division
5463 Dunham Rd.
Maple Heights, OH 44137
USA

phone: 216 662 3100
fax: 216 663 7504

Aluminum and alloys; Beryllium and alloys; Copper and alloys; Indium; Lead and alloys; Tin and alloys; Zinc and alloys

Alumix S.P.A.

Piazza Marconi 25
I-00144 Rome
Italy

phone: 39 06 549 031
fax: 39 06 549 03360

Aluminum and alloys

OTHER LOCATIONS

Alumix GmbH
Deutschland Ingelstrasse 24
4000 Duessaldorf 1
Germany
phone: 49 211 4930803
fax: 49 211 4910864

Alumix Ltd.
10 Eghams Court
Boston Drive
Bourne End, Bucks SL8 5YS
England
phone: 44 6285 30552
fax: 44 6285 30266

Alumix GmbH
Johannes Vermeerplein 5
1071 Amsterdam
Netherlands
phone: 31 20 736886
fax: 31 20 736126

Alumix S.A.
61 Rue de l'Arcade
75008 Paris
France
phone: 33 1 45224849
fax: 33 1 45223006

(continued)

Alumix S.P.A. (cont'd)

Alumix S.A.
Calle Santa Engracia 141
28008 Madrid
Spain
phone: 34 1 5341135
fax: 34 1 5340626

Alumix S.P.A.
Via S. Martino 60
20017 Rho (Milano)
Italy
phone: 39 2 933361
fax: 39 2 93500852

Alumix S.P.A.
Via Caldera 21
20153 Milan
Italy
phone: 39 2 452821
fax: 39 2 45282265

Alumix S.P.A.
Viale deil'Elettricita 19/B
30175 Porto Marghera (Venezia)
Italy
phone: 39 41 5380108
fax: 39 41 5381424

Alumix S.P.A.
09010 Portoscuso (Cagliari)
Italy
phone: 39 0781 5031
fax: 39 0781

Alusuisse-Lonza Group

Trading AG
Max Hogger-Str. 6
Postfach 1149
CH-8048 Zurich
Switzerland
phone: 011 41 01 431 1
fax: 011 41 01 433 0

Aluminum and alloys

Amalga Composites, Inc.

10600 West Mitchell St.
West Allis, WI 53214
USA

phone: 800 394 9545
fax: 414 453 9561

Epoxies

Ambo-Stahl Gesellschaft

Gerhard Sevenich GmbH & Co.
Oskar Jager Strasse 155
D-5000 Koln 30
Germany

phone: 011 49 221 545 915
fax: 011 49 221 545 919

Alloy steels, mill shapes; Stainless steels, mill shapes; Tool steels

American Chain & Cable

Page Aluminized Steel Corp.
100 Monongahela St.
Monessen, PA 15062
USA

phone: 412 684 4000

Stainless steels; Tool steels; Copper and alloys; Zinc and alloys

American Crucible Products

a Division of Johnson Metall, Inc.
1305 Oberlin Ave.
Lorain, OH 44052
USA

phone: 216 245 6826
fax: 216 244 0414

Copper and alloys; Lead and alloys

American Cyanamid Co.

21444 Golden Triangle Road
Saugus, CA 91350
USA

phone: 805 259 1415
fax: 805 254 2779

Epoxies; Phenolics; Polyesters, thermoset; Polyimides; Fibers

American Saw & Mfg. Co.

301 Chestnut Street
East Longmeadow, MA 01028
USA

phone: 800 628 3030
fax: 413 525 2336

Stainless steels

American Smelting & Refining Co.

ASARCO,Inc.
180 Maiden Lane
New York, NY 10038
USA

phone: 212 510 2000
fax: 212 510 1835

Aluminum and alloys; Cadmium, chromium; Copper and alloys; Indium; Lead and alloys; Tin and alloys; Zinc and alloys

American Tank & Fabricating Co.

12314 Elmwood Ave.
Cleveland, OH 44111
USA

phone: 216 252 1500
fax: 216 252 4871

Alloy steels, sheet, strip

Amoco Performance Products, Inc.

4500 McGinnis Ferry Road
Alpharetta, GA 30202
USA

phone: 404 772 8200
fax: 404 772 8753

Polyamides; Polyamide-imides; Polyesters, thermoplastic; Polyketones; Polysulfones

AMP Inc.

Flexible Film Sensors
PO Box 799
Valley Forge, PA 19482-0799
USA

phone: 215 666 3500
fax: 215 666 3509

Fluoropolymers, chloropolymers

Ampal, Inc.

P.O. Box 124
Palmerton, PA 18071
USA

phone: 908 782 5454
fax: 908 782 3489

Aluminum and alloys

Ampco Metal

P.O. Box 2004
1745 South 38th Street
Milwaukee, WI 53201
USA

phone: 414 645 3750
fax: 414 645 3225

Copper and alloys

Ancast, Inc.

3194 Townline Road
Sodus, MI 49126
USA

phone: 616 927 1985

Stainless steels; Nickel, alloys, superalloys

Anor, Acieries et Forges

40 Rue de Marechal Foch
F59186 Anor
France

fax: 33 27 59 6164

Stainless steels; Cast iron; Tool steels; Nickel, alloys, superalloys

Anval, Inc.

301 Rt. 17, North
Suite 800
Rutherford, NJ 07070
USA

phone: 201 939 1065
fax: 201 939 1608

Ferrous welding, hard facing electrodes

Apollo Metals Ltd.

1001 14th Avenue
P.O. Box 4045
Bethlehem, PA 18018
USA

phone: 800 338 0199
fax: 215 867 7594

Stainless steels

Applied Composites

333 North Sixth St.
St. Charles, IL 60174
USA

phone: 708 584 3130
fax: 708 584 0659

Polyesters, thermoset

ARC Mondial GmbH

Oberloh 8
D-5460
Linz/Rhein
Germany

phone: 49 02644 6311 2
fax: 49 02644 1404

Stainless steels, cast

Arco Chemical Co.

Subsidiary of Atlantic Richfield
3801 West Chester Pike
Newtown Square, PA 19073
USA

phone: 215 359 2000
fax: 215 359 2841

Polyolefins; Polystyrenes

Arcos Alloys

P.O. Box 308
One Arcos Drive
Mount Carmel, PA 17851-0308
USA

phone: 800 233 8460
fax: 717 339 5206

Ferrous welding, hard facing electrodes; Nickel, alloys, superalloys

Aristech Chemical Corp.

7350 Empire State Dr.
Florence, KY 41042
USA

phone: 800 755 6353
fax: 606 283 6497

Acrylics; Polyallomers

Armco Inc.

300 Interpace Parkway
Parsippany, NJ 07054-0324
USA

phone: 201 316 5276
fax: 201 316 5203

Stainless steels; Stainless steels, sheet, strip; Ferrous electronic materials

OTHER LOCATIONS

Armco Advanced Materials Corp.
Stainless Sheet & Strip Products
PO Box 832
Butler, PA 16003-0832
USA
phone: 412 284 2000
fax: 412 284 2874

Armco Stainless and Alloy Products
3501 East Biddle Street
Baltimore, MD 21213-4097
USA
phone: 800 638 1452
fax: 410 563 5933

Armco Steel Company

703 Curtis Street
Middletown, OH 45043
USA

phone: 513 425 5876
fax: 513 425 5613

Carbon steels, sheet, strip; Carbon steels, plate

Arnold Engineering Co.

P.O. Box G
300 N. West St.
Marengo, IL 60152
USA

phone: 815 568 2000
fax: 815 568 2228

Ferrous electronic materials

Ashley Polymers Inc.

5114 Fort Hamilton Pkwy.
Brooklyn, NY 11219
USA

phone: 718 851 8111

Polyamides

Associated Spring Co.

Barnes Group Inc.
18 Main St.
Bristol, CT 06010
USA

phone: 800 528 7795
fax: 203 582 1358

Alloy steels; Stainless steels; Copper and alloys

OTHER LOCATIONS

Associated Spring Co.
Gardena, CA
USA
phone: 310 323 4833
fax: 310 327 0379

Associated Spring Co.
Glen Ellyn, IL
USA
phone: 708 858 1030
fax: 708 858 6150

(continued)

Associated Spring Co. (cont'd)

Associated Spring Co.
Burlington, Ontario
Canada

phone: 416 335 6688
fax: 416 336 1336

Associated Steel Corp.

P.O. Box 28335
18200 Miles Ave.
Cleveland, OH 44128
USA

phone: 216 475 8000

Tool steels

Astro Metallurgical

Div. of Harsco Corp.
P.O. Drawer 520
3225 Lincoln Way West
Wooster, OH 44691
USA

phone: 216 264 8639
fax: 216 264 4316

Titanium and alloys

OTHER LOCATION

Astro Metallurgical Inc.
Northwoods Industrial Park West
12231 F.M. 529
Houston, TX 77041
USA

Atlantic Casting & Engineering Corp.

810 Bloomfield Ave.
Clifton, NJ 07012
USA

phone: 201 779 2450

Aluminum and alloys; Copper and alloys

Atlantic Steel Co.

P.O. Box 1714
Atlanta, GA 30301
USA

phone: 404 897 4505
fax: 404 897 4623

Carbon steels, mill shapes; Alloy steels, mill shapes; Alloy steels, plate

Atlas Brass Foundry

10729 Forest Street
Santa Fe Springs, CA 90670
USA

phone: 310 903 5670
fax: 310 903 5699

Copper and alloys

Atlas Foundry & Machine Co.

3021 South Wilkeson Street
Tacoma, WA 98409-7857
USA

phone: 206 475 4600
fax: 206 473 8710

Stainless steels, cast

Atomergic Chemetals Corp.

222 Sherwood Ave.
Farmingdale, NY 11735-1527
USA

phone: 516 694 9000
fax: 516 694 9177

Aluminum and alloys; Beryllium and alloys; Cadmium, chromium; Cobalt, alloys, superalloys; Copper and alloys; Indium; Lead and alloys; Magnesium and alloys; Manganese; Nickel, alloys, superalloys; Nuclear, heavy metals, thorium; Precious metals; Refractory metals; Tin and alloys; Titanium and alloys; Vanadium; Zinc and alloys; Zirconium; Carbon, graphite, diamond

Atrax Co.

Wallace Murray Corp.
205 N. 13th Street
P.O. Box 9
Rogers, AK 72757-0009
USA

phone: 800 633 5994
fax: 501 636 4638

Carbides

Aubert et Duval

Acierie des Ancizes
41, rue de Villiers
BP 120
92202 Neuilly-sur-Se
France

phone: 33 1 40 88 20 09
fax: 33 1 47 57 69 39

Alloy steels; Alloy steels; Alloy steels, mill shapes; Stainless steels; Iron-base superalloys; Tool steels; Cobalt, alloys, superalloys; Nickel, alloys, superalloys

OTHER LOCATIONS

Aubert & Duval
Usine des Ancizes
63770 Les Ancizes Co
France
phone: 33 73 86 04 21
fax: 33 73 86 79 39

Aubert et Duval
22 rue Henri Vuillemin
92230 Gennevilliers
France
phone: 33 1 47 90 61 34
fax: 33 1 40 86 30 17

Aurora Industries

Aurora Metals Division
1995 Greenfield Road
Montgomery, IL 60538
USA

phone: 708 844 4900
fax: 708 844 6839

Copper and alloys

Ausimont USA Inc.

44 Whippany Rd.
Morristown, NJ 07960-1838
USA

phone: 800 221 0553
fax: 201 292 0886

Fluoropolymers, chloropolymers

Avesta Sheffield Group

P.O. Box 370
549 W. St. Rd. 38
New Castle, IN 47362
USA

phone: 317 529 0120

Stainless steels, mill shapes; Stainless steels, sheet, strip; Stainless steels, plate

OTHER LOCATIONS

Avesta ABE Inc.
1800 J MacLeod Drive
Lawrenceville, GA 30243
USA
phone: 404 339 9556

Avesta Sheffield A/S
Box 50
DK-2625 Vallensbaek
Denmark
phone: 45 43 62 45 33

Avesta Sheffield (SEA) Pte. Ltd.
156 Paya Lebar Road
Singapore 1440
Singapore
phone: 65 741 7900

Avesta Sheffield A/S
P.O. Box 41 Ljan
N-1113 Oslo
Norway
phone: 47 22 62 99 00

(continued)

Avesta Sheffield Group (cont'd)

Avesta Sheffield AB
Vasagatan 8-10
Box 16377
S10327 Stockholm
Sweden
phone: 46 8 6133600
fax: 46 8 208481

Avesta Sheffield B.V.
P.O. Box 41146
NL-1009 EC Amsterdam
Netherlands
phone: 31 0 20 665 6061

Avesta Sheffield Do Brasil Ltda.
P.O. Box 1185
01059-970 Sao Paulo-SP
Brazil
phone: 55 011 284 5081

Avesta Sheffield GesmbH
P.O. Box 31
A-2345 Brunn am Gebirge
Austria
phone: 43 0 2236 33500

Avesta Sheffield GmbH
P.O. Box 254
D-4156 Willich
Germany
phone: 49 0 2154 4960

Avesta Sheffield Inc.
2140 Meadowpine Blvd.
Mississauga, Ontario L5N 6H6
Canada
phone: 416 567 9900

Avesta Sheffield Ireland Ltd.
57 Fairview Strand
Fairview, Dublin 3
Ireland
phone: 353 0 1 368600

Avesta Sheffield Ltd.
P.O. Box 161
Shepcote Lane
Sheffield S91 TR
Great Britain
phone: 44 0742 443311
fax: 44 0742 448280

Avesta Sheffield N.V.
Luchthavenlaan 25-27
B-1800 Vilvoorde
Belgium
phone: 32 0 2 252 0607

Avesta Sheffield Oy
P.O. Box 130
SF-04301 Tuusula
Finland
phone: 358 9 0 273 991

Avesta Sheffield Pipe Co.
1101 North Main Street
Wildwood, FL 34785
USA
phone: 904 748 1313

Avesta Sheffield S.A.
46/52 rue Albert
F-75013 Paris
France
phone: 33 1 44 06 83 00

(continued)

Avesta Sheffield Group (cont'd)

Avesta Sheffield S.A.
Avenida de Burgos, 16E
E-28036 Madrid
Spain
phone: 34 9 1 302 7444

Avesta Sheffield S.A. de C.V.
Apartado Postal 21-313
Col Del Carmen
Coyoacan 04100
Mexico D.F.
Mexico
phone: 52 5 659 7847

Avesta Sheffield SpA
P.O. Box 16001
I-20160 Milan
Italy
phone: 39 0 2 69651

Avesta Sheffield sro
Zahostice 39
CS 39155 Chynov
Czechoslovakia
phone: 42 0 361 90 114

Avesta Sheffield Inc.
Woodfield Corporate Center
425 N. Martingale Road
Suite 2000
Schaumburg, IL 60173
USA
phone: 800 833 8703
fax: 800 545 8617

Avesta Welding Products Inc.
3176 Abbott Road
Orchard Park, NY 14127
USA
phone: 716 827 4400

Energy Steel Products Corp.
10002 Windfern
Houston, TX 77064
USA
phone: 713 890 6666

Fox Metals, Inc.
10002 Windfern
Houston, TX 77064
USA
phone: 713 890 6666

T.S. Alloys Inc.
5022 Ashley Court
Houston, TX 77041
USA
phone: 713 896 1200

Axel Johnson Metals Inc.

215 Welsh Pool Road
Exton, PA 19341
USA

phone: 215 363 0330
fax: 215 524 1567

Titanium and alloys

Babcock & Wilcox

Power Generation Group
20 South Van Buren Ave.
PO Box 351
Barberton, OH 44203-0351
USA

phone: 216 753 4511
fax: 216 860 1886

Stainless steels; Tool steels

BALCO

Bahrain Saudi Aluminum Marketing Co.
P.O. Box 20079
Bahrain

phone: 966 532626
fax: 966 532727

Aluminum and alloys

Bamberger Polymers, Inc.

1983 Marcus Ave.
Lake Success, NY 11042
USA

phone: 516 328 2772
fax: 516 326 1005

Polyesters, thermoplastic

Barker & Allen Ltd.

Nickel Silver Works
Heath Street South
Springhill
Birmingham B18 7PZ
England

phone: 44 021 452 1199
fax: 44 021 456 1595

Copper and alloys

Barworth Flockton Ltd.

Ecclesfield
Sheffield, SYorks S30 3XH
England

phone: 44 0742 468291
fax: 44 0742 454230

Tool steels

BASF Corp.

Plastic Materials
100 Cherry Hill Road
Parsippany, NJ 07054
USA

phone: 800 227 3446
fax: 201 335 6730

Acetals; Acrylics; Polyamides; Polyesters, thermoplastic; Polyesters, thermoset; Polyketones; Polyolefins; Polyphenylenes; Polyurethanes; Polystyrenes; Polysulfones; Polyvinyl chlorides; Thermoplastic elastomers

Bearium Metals Corp.

1170 Chili Ave.
Rochester, NY 14624
USA

phone: 716 235 5360
fax: 716 235 7716

Tin and alloys

Beaumont Birch Co.

3900 River Road
P.O. Box 599
Pennsauken, NJ 08110
USA

phone: 609 663 6440
fax: 609 663 2732

Stainless steels

Becker Bros. Carbon Corp.

3450 S. Laramie Ave.
Cicero, IL 60650
USA

phone: 312 242 1260
fax: 312 242 1023

Ferrous welding, hard facing electrodes

Belmont Metals, Inc.

330 Belmont Ave.
Brooklyn, NY 11207
USA

phone: 718 342 4900
fax: 718 342 0175

Aluminum and alloys; Copper and alloys; Nickel, alloys, superalloys

Bendix Steel Co.

1612 Enterprise Pkwy.
Twinsburg, OH 44087
USA

phone: 216 425 3600

Tool steels

Bethlehem Steel Corp.

701 E. Third Street
Bethlehem, PA 18016-7699
USA

phone: 215 694 2424

Carbon steels; Carbon steels, mill shapes; Carbon steels, sheet, strip; Carbon steels, plate; Carbon steels, forged; Alloy steels; Alloy steels, mill shapes; Alloy steels, sheet, strip; Alloy steels, sheet, strip; Alloy steels, plate; Alloy steels, forged; Stainless steels; Tool steels

OTHER LOCATIONS

Bethlehem Steel Corp.
5111 North Point Blvd.
Sparrows Point, MD 21219-1014
USA
phone: 800 521 4789
fax: 800 626 1279

Bethlehem Steel Corp.
501 E. Third St.
Room 426
Bethlehem, PA 18016-7699
USA
phone: 800 633 0482
fax: 800 535 9530

BGH Edelstahl GmbH

Industriestrasse 9
Postfach 21 01 63
5900 Siegen
Germany

phone: 49 0271 701 0
fax: 49 0271 701 300

Alloy steels, mill shapes; Alloy steels, forged; Stainless steels, mill shapes; Stainless steels, forged; Iron-base superalloys

(continued)

BGH Edelstahl GmbH (cont'd)

OTHER LOCATIONS

BGH Edelstahl Freital GmbH
Postfach 1-64
01691 Freital
Germany
phone: 49 0351 646 0
fax: 49 0351 646 2048

BGH Edelstahl Lugau GmbH
Postfach 27
09382 Lugau
Germany
phone: 49 037295 29 0
fax: 49 037295 29 570

BGH Edelstahl Leuna GmbH
Leuna-Werke G 17
06236 Leuna
Germany
phone: 49 0345 5580266
fax: 49 0345 5580268

Billiton International Metals BV

PO Box 436
NL 2260 AK Leidschen
Netherlands

phone: 31 070 317 3911
fax: 31 070 317 2700

Ferroalloys, pure iron; Tool steels; Aluminum and alloys; Cadmium, chromium; Lead and alloys; Precious metals; Refractory metals; Tin and alloys; Zinc and alloys; High alumina and "cermets"

(continued)

Billiton International Metals BV

OTHER LOCATIONS

Billiton Metais SA
Praia de Botafogo 370, 4th Floor
Caixa Postal 62050
22250-040 Rio de Janeiro
Brazil
phone: 55 21 559 7000
fax: 55 21 559 7599

Billiton Witmetaal BV
P.O. Box 5018
1410 AA Naarden
Netherlands
phone: 31 2159 55411
fax: 31 2159 48451

Billiton Witmetaal France SA
8 Rue Francois Arago
BP 141
92185 Antony Cedex
France
phone: 33 1 4237 1400
fax: 33 1 4237 1610

Billiton Witmetaal GmbH
Duisburger Strasse 44
Postfach 61 01 30
8500 Nurnberg 60
Germany
phone: 49 911 996 300
fax: 49 911 996 3050

Billiton Zink BV
P.O. Box 2135
6020 AC Budel
Netherlands
phone: 31 4950 12700
fax: 31 4950 18225

British Lead Mills
Peartree Lane
Welwyn Garden City
Hertfordshire AL7 3UB
England
phone: 44 707 324595
fax: 44 707 328941

Budelco BV
P.O. Box 2001
6020 AA Budel-Dorplein
Netherlands
phone: 31 4950 12911
fax: 31 4950 18285

Hollandse Metallurgische Industrie Billiton BV
P.O. Box 38
6800 LH Arnhem
Netherlands
phone: 31 85 654501
fax: 31 85 621995

(continued)

Billiton International Metals BV (cont'd)

Le Plomb Francais SA
Zone Industrielle Le bois Chev
60190 Estrees Saint
France
phone: 33 4441 3131/0108
fax: 33 4441 0090

Thailand Smelting and Refining Company Ltd. (Thaisarco)
CPO Box 345
Bankok 10501
Thailand
phone: 66 2 249 0491
fax: 66 2 249 8335

Metals Representative, Shell Pacific Enterprises Ltd.
Korea Branch
P.O. Box 396
Gwanghwamoon, Seoul
South Korea
phone: 82 2 363 0041
fax: 82 2 393 6196

Birmingham Battery & Metal Co. Ltd.

Bristol Road
Selly Oak
Birmingham B29 6AB
UK

phone: 44 021 472 1151
fax: 44 021 414 1168

Copper and alloys

Black Drill Company, Inc.

30132 Lakeland Blvd.
Cleveland, OH 44092
USA

phone: 216 943 1224
fax: 216 943 1275

Tool steels

Black-Clawson Co.

605 Clark St.
Middletown, OH 45042
USA

phone: 513 424 7400

Tool steels

Blackalloy Co. of America, Inc.

10 Vesper St.
P.O. Box 2157
Paterson, NJ 07509
USA

phone: 201 742 1234
fax: 201 742 1205

Stainless steels; Stainless steels, cast; Carbides

Blaw Knox Rolls

98 Peninsula Street
Wheeling, W. Va. 26003
USA

phone: 304 234 9600

Cast iron

Bliss & Laughlin Steel Co.

281 East 155th Street
Harvey, IL 60426
USA

phone: 312 264 1800
fax: 708 333 3786

Alloy steels, mill shapes

Bohler Edelstahl GmbH

Mariazeller Strasse 25
Postfach 96
A-8605 Kapfenberg, Osterreich
Austria

phone: 43 3862 20 0
fax: 43 3862 20 7589

Carbon steels, mill shapes; Alloy steels, mill shapes; Stainless steels; Stainless steels, mill shapes; Tool steels; Ferrous electronic materials; Aluminum and alloys; Cobalt, alloys, superalloys; Copper and alloys; Nickel, alloys, superalloys; Titanium and alloys; Carbides

Borden Packaging

Div. of Borden,Inc.
Holt Road
North Andover, MA 01845
USA

phone: 617 682 0701
fax: 800 345 4049

Polyolefins

Borg-Warner Chemicals, Inc.

International Center
Parkersburg, WV 26102
USA

phone: 304 424 5411

Polyphenylenes; Polystyrenes

BP Chemical

Commercial Composites
3535 Latonia Ave.
P.O. Box 15159
Ft Wright, KY 41015-0159
USA

phone: 606 491 5805
fax: 606 292 7404

Polyesters, thermoset

Brighton Electric Steel Casting Co.

510 45th Street
P.O. Box 206
Beaver Falls, PA 15010
USA

phone: 412 846 7377
fax: 412 846 9124

Tool steels

Brillion Iron Works

200 Park Ave.
Brillion, WI 54110
USA

phone: 414 756 2121
fax: 414 756 3550

Cast iron

Bristol Aluminum Co. (All-Lite Co.)

5514 Emilie Road
Levittown, PA 19057
USA

phone: 215 946 3160

Aluminum and alloys

British Alcan Aluminium plc

Chalfont Park
Gerrards Cross
Bucks SL9 0QB
England

phone: 44 753 887373
fax: 44 753 889667

Aluminum and alloys

British Driver-Harris Co.

308 Middlesex St.
Harrison, NJ 07029
USA

phone: 201 483 4802
fax: 201 483 4806

Ferrous welding, hard facing electrodes; Ferrous electronic materials;Copper and alloys; Nickel, alloys, superalloys

OTHER LOCATIONS

British Driver-Harris Co.
Pepper Road
Hazel Grove
Stockport, Cheshire SK7 5BP
U.K.
phone: 44 61 456 3131
fax: 44 61 456 1867

Driver-Harris Australia Pty. Ltd.
Childs Road
Epping, Victoria 3076
Australia
phone: 61 3 401 1222
fax: 61 3 401 3541

Driver-Harris Iberica
apartado 49.006
Madrid 12
Spain
phone: 34 1 776 5181
fax: 34 1 776 2956

Driver-Harris Italiana S.P.A.
Via Ospiate 51
20017 Terrazzano Di Rho
Milano
Italy
phone: 39 2 930 6671
fax: 39 2 930 9445

Driver-Harris Japan Co. Ltd.
13-27 Takakuracho
1 Chome Miyakojima-Ku
Osaka
Japan
phone: 81 6 922 2966
fax: 81 6 922 0332

Driver-Harris S.A.
Rue De Buchelay
Boite Postale 1062
Mantes-La-Jolie 78202
France
phone: 33 1 3094 4800
fax: 33 1 3476 8276

Irish Driver-Harris Co. Ltd.
5A, Ballymount Trading Estate
Lower Ballymount Road
Walkinstown
Dublin 12
Ireland
phone: 353 1 506935
fax: 353 1 503330

British Oxygen Co., Ltd.

Deerpark
London SW19 3UF
England

Nickel, alloys, superalloys

British Rolling Mills Ltd.

Brymill Steel Works
Bloomfield Road
Tipton
West Midlands DY4 9EF
England

phone: 44 021 557 3939
fax: 44 021 557 6047

Alloy steels, mill shapes; Tool steels

British Steel plc

9 Albert Enbankment
London SE1 7SN
England

phone: 44 71 735 7654
fax: 44 71 587 1142

Carbon steels, mill shapes; Carbon steels, sheet, strip; Carbon steels, plate; Alloy steels, mill shapes; Alloy steels, sheet, strip; Alloy steels, plate; Stainless steels; Stainless steels, mill shapes; Stainless steels, sheet, strip; Stainless steels, plate; Iron-base superalloys; Ferrous welding, hard facing electrodes; Tool steels

Brontel Bearing Bronze Company

9314 Elizabeth Ave.
Cleveland, OH 44105
USA

phone: 216 641 6520
fax: 216 641 0570

Copper and alloys

Brush Wellman

Engineered Materials
17876 St. Clair Ave.
Cleveland, OH 44110
USA

phone: 800 321 2076
fax: 216 383 4091

Beryllium and alloys

OTHER LOCATIONS

Brush Wellman (Japan) Ltd.
Dai-ichi Marusan Building
9, Kanda Jimbocho 3-chrome
Chiyoda-ku
Tokyo 101
Japan
phone: 81 03 3230 2961
fax: 81 03 3230 2908

Brush Wellman GmbH
Motorstrasse 34
D-7000 Stuttgart 31
Germany
phone: 49 0711 83093 0
fax: 49 0711 83338 22

Brush Wellman Inc.
18720 Crenshaw Blvd.
Torrance, CA 90504
USA
phone: 800 421 2733
fax: 310 715 2316

Brush Wellman Inc.
606 Lamont Road
Elmhurst, IL 60126
USA
phone: 800 323 2438
fax: 708 832 9657

(continued)

Brush Wellman (cont'd)

Brush Wellman Inc.
27555 College Park Drive
Warren, MI 48093
USA
phone: 800 521 8800
fax: 313 772 2472

Brush Wellman Ltd.
4-5 Ely Road
Theale Commercial Estate
Theale
Reading, Berkshire RG7 4BQ
England
phone: 44 0734 303733
fax: 44 0734 303635

Budd Company

Plastics Div.
32055 Edward Ave.
Madison Hts., MI 48071
USA

phone: 313 588 3200

Epoxies; Polyesters, thermoset

Buffalo Wire Works Co.

P.O. Box 129
Buffalo, NY 14240
USA

phone: 716 826 4666

Tool steels

Bulk Molding Compounds Inc.

3N497 N. 17th St.
St. Charles, IL 60174
USA

phone: 708 377 1065
fax: 708 377 7395

Polyesters, thermoset

Bull Moose Tube Co.

1819 Clarkson Road
Suite 100
Chesterfield, MO 63017
USA

phone: 314 537 2600
fax: 314 537 2645

Carbon steels, mill shapes

Bundy Corp

Bundy Tubing Div.
12345 E. 9 Mile Road
Warren, MI 48090
USA

phone: 313 758 4511
fax: 313 758 5167

Carbon steels, mill shapes

Bunting Bearings Corp.

200 Van Buren St.
P.O. Box 37
Delta, OH 43515
USA

phone: 419 822 3456
fax: 419 822 3372

Copper and alloys

OTHER LOCATIONS

Bunting Bearings Corp.
4252 E. Kilgore Road
Kalamazoo, MI 49002
USA
phone: 616 345 8691
fax: 616 345 0931

Bunting Bearings Corp.
1001 Holland Park Blvd.
P.O. Box 729
Holland, OH 43528
USA
phone: 419 866 7000
fax: 419 866 0653

Bunting Bearings Corp.
153 E. Fifth St.
P.O. Box 1053
Mansfield, OH 44902
USA
phone: 419 522 3323
fax: 419 522 6011

Bunting Bearings Corp.
8935 Market St.
Houston, TX 77029
USA
phone: 713 672 7669
fax: 713 672 0462

Bunting Bearings Corp.
12629 Hidden Creek Way
Cerritos, CA 90701
USA
phone: 310 404 3721
fax: 310 926 2218

Burndy Corp.

51 Richards Ave.
Norwalk, CT 06856
USA

phone: 203 838 4444
fax: 203 852 6227

Cast iron; Tool steels; Copper and alloys

Busch-Jager Ludenscheider Metallwerke

Altenaerstrasse 109
P.O. Box 1360
D 5880 Ludenscheid
Germany

phone: 49 02351 181 0
fax: 49 02351 181 200

Aluminum and alloys

Cadman, A.W., Manufacturing Co.

2816 Smallman St.
Pittsburgh, PA 15222
USA

phone: 412 281 6683
fax: 412 281 8643

Lead and alloys; Tin and alloys

Calsak Corporation

P.O. Box 9036
Compton, CA 90224
USA

phone: 310 637 2000
fax: 310 637 3696

Acrylics; Polyvinyl chlorides

Calumet Steel Company

Box 217
317 East 11th St.
Chicago Hts., IL 60411
USA

phone: 708 757 7300
fax: 708 757 7901

Carbon steels, mill shapes; Alloy steels, mill shapes

Cambridge Inc.

P.O. Box 399
105 Goodwill Road
Cambridge, MD 21613
USA

phone: 410 228 3000
fax: 410 228 6752

Carbon steels; Stainless steels

Camden Wire Co. Inc.

A Subsidiary of Oneida Ltd.
12 Masonic Ave.
Camden, NY 13316
USA

phone: 315 245 2000
fax: 315 245 1916

Copper and alloys

Cannon-Muskegon Corp.

Box 506
Muskegon, MI 49443-0506
USA

phone: 616 755 1681
fax: 616 755 4975

Stainless steels; Iron-base superalloys; Cobalt, alloys, superalloys; Nickel, alloys, superalloys

OTHER LOCATION
Cannon-Muskegon Corp.
2875 Lincoln Street
Muskegon, MI 49441
USA

Capitol Castings Inc.

P.O. Box 27328
Tempe, AZ 85285-7328
USA

phone: 602 839 3000
fax: 602 730 0649

Alloy steels, cast; Stainless steels, cast

OTHER LOCATION
Capitol Castings Inc.
5857 S. Cyrene
Tempe, AZ 85283
USA

phone: 602 839 3000
fax: 602 730 0649

Carbidie

P.O. Box 509
Arona Rd.
Irwin, PA 15642-0509
USA

phone: 800 245 6830
fax: 412 864 2443

Carbides

OTHER LOCATION
Carbidie Asia Pacific Pte. Ltd.
No. 6 Jalan Ampas
#01-03, #02-03
Singapore 1232
Singapore

phone: 65 2543601
fax: 65 2560261

Carboloy Inc.

P.O. Box 330237
Detroit, MI 48232
USA

phone: 800 832 8326
fax: 313 497 5627

Iron-base superalloys; Cobalt, alloys, superalloys; Nickel, alloys, superalloys; Refractory metals; High alumina and "cermets"

OTHER LOCATIONS

Dormer Tools Canada Ltd.
6835 Century Avenue
Mississauga, Ontario L5N 2L2
Canada
phone: 905 542 3000
fax: 905 542 7000

Outils Dormer Ltee
2535 Boul. Chomedey
Laval, Quebec H7T 2R2
Canada
phone: 514 682 4300
fax: 514 682 9775

Carborundum Company

P.O. Box 156
Niagara Falls, NY 14302
USA

phone: 716 278 2000

Electrical ceramics; Refractory ceramics; High alumina and "cermets"; Carbides; Nitrides

OTHER LOCATIONS

Carborundum Company
345 Third Street
Niagra Falls, NY 14303
USA

Carborundum Company
Boron Nitride Division
Audubon Industrial Park
Amherst, NY 14228-2027
USA
phone: 716 691 2051
fax: 716 691 2090

(continued)

Carborundum Company (cont'd)

Carborundum Company
Carborundum Specialty Products
P.O. Box 808
Niagra Falls, NY 14302-0808
USA
phone: 716 278 2480
fax: 716 278 2963

Carborundum Company
Ekonol Poyester Resins
P.O. Box 2083
Niagra Falls, NY 14302
USA
phone: 716 278 6341
fax: 716 278 2900

Carborundum Company
Fibers Division
P.O. Box 808
Niagra Falls, NY 14302-0808
USA
phone: 716 278 6221
fax: 716 278 4962

Carborundum Company
Microelectronics Division
10409 South 50th Place
Phoenix, AZ 85044
USA
phone: 602 496 5000
fax: 602 496 5060

Carborundum Company
Monofrax Refractories Division
501 New York Avenue
Falconer, NY 14733
USA
phone: 716 483 7200
fax: 716 665 2478

Carborundum Company
Performance Refractories Division
End Crows Mill Road West
Keasbey, NJ 08832
USA
phone: 908 417 3000
fax: 908 417 3055

Carborundum Company
Structural Ceramics Division
1625 Buffalo Avenue
Niagra Falls, NY 14303-1597
USA
phone: 716 278 6233
fax: 716 278 3909

Carborundum Company
Substrates Division
2050 Cory Road
Sanborn, NY 14132-9633
USA
phone: 716 731 9244
fax: 716 731 9256

Carlson, G.O., Inc.

P.O. Box 526
350 Marshallton Rd.
Thorndale, PA 19372-0526
USA

phone: 800 345 8070
fax: 215 383 3429

Stainless steels, mill shapes; Stainless steels, plate; Nickel, alloys, superalloys

Carmet Company

1 Tungsten Way
Duncan, SC 29334
USA

phone: 803 879 7621
fax: 803 877 7425

Carbides

Carondelet Corp.

P.O. Box 769
8600 Commercial Blvd.
Pevely, MO 63070-0769
USA

phone: 314 479 4499
fax: 314 479 3399

Stainless steels, cast

Carpenter Technology Corp.

P.O. Box 14662
Reading, PA 19612-4662
USA

phone: 215 208 2000
fax: 215 371 2858

Alloy steels, mill shapes; Stainless steels; Stainless steels, mill shapes; Stainless steels, sheet, strip; Iron-base superalloys; Tool steels; Ferrous electronic materials; Nickel, alloys, superalloys

OTHER LOCATIONS

Carpenter Technology (Canada) Ltd.
7464 Tranmere Drive
Unit 2
Mississauga, Ontario L5S 1K4
Canada
phone: 416 672 9150
fax: 416 672 9156

Carpenter Technology (UK) Ltd.
Unit 48, Eagle Road
North Moons Moat
Redditch, Worcestershire B98 9HD
England
phone: 44 0527 68112
fax: 44 0527 66769

Carpenter Technology (Deutschland) GmbH
Waldenbucher Strasse 22
D-71065 Sindelfingen
Germany
phone: 49 7031 875277
fax: 49 7031 870353

Carpenter Technology Corp.
101 W. Bern Street
Reading, PA 19601
USA

Carpenter Technology (Europe) S.A.
207-209 Boulevard Auguste Reyers
1040 Brussels
Belgium
phone: 32 2 7339632
fax: 32 2 7335200

Carpenter Technology Corp.
Snapfinger Woods Industrial Park
5355 Morse Drive
Decatur, GA 30035-3810
USA
phone: 404 987 3000
fax: 404 987 3062

(continued)

Carpenter Technology Corp. (cont'd)

Carpenter Technology Corp.
234 Forbes Blvd.
P.O. Box 620
Mansfield, MA 02048-0620
USA
phone: 508 339 5161
fax: 508 261 8880

Carpenter Technology Corp.
4501 James Place
Melrose Park, IL 60160-1080
USA
phone: 708 681 3500
fax: 708 681 1006

Carpenter Technology Corp.
680 Redna Terrace
P.O. Box 15507
Cincinnati, OH 45215-0507
USA
phone: 513 771 0250
fax: 513 772 6065

Carpenter Technology Corp.
4901 West 150th Street
Cleveland, OH 44135-3398
USA
phone: 216 267 6700
fax: 216 267 7989

Carpenter Technology Corp.
2605 Freewood Drive
Dallas, TX 75354-0065
USA
phone: 214 352 1771
fax: 214 352 6187

Carpenter Technology Corp.
2012 Springboro West
Moraine, OH 45439-1648
USA
phone: 513 224 7456
fax: 513 772 6065

Carpenter Technology Corp.
Executive Park East
446 Executive Drive
Troy, MI 48083-7988
USA
phone: 313 588 5300
fax: 313 588 6466

Carpenter Technology Corp.
847 West River Center N.E.
Comstock Park, MI 49321
USA
phone: 616 785 9922

Carpenter Technology Corp.
277 New Park Avenue
P.O. Box 2419
Hartford, CT 06146-2419
USA
phone: 203 233 7581
fax: 203 231 2131

Carpenter Technology Corp.
610 Lockhaven Drive
Houston, TX 77073-5505
USA
phone: 713 821 0987
fax: 713 821 9729

(continued)

Carpenter Technology Corp. (cont'd)

Carpenter Technology Corp.
5905 West 74th Street
P.O. Box 681338
Indianapolis, IN 46268-7338
USA
phone: 317 297 3300
fax: 317 293 2570

Carpenter Technology Corp.
4423 East District Boulevard
P.O. Box 58880
Los Angeles, CA 90058-0880
USA
phone: 213 587 9131
fax: 313 587 8145

Carpenter Technology Corp.
3070 Connahbrook Drive
P.O. Box 16620
Memphis, TN 38186-0620
USA
phone: 901 332 8332
fax: 901 332 3643

Carpenter Technology Corp.
4330 West Green Tree Road
P.O. Box 23239
Milwaukee, WI 53223-0239
USA
phone: 414 353 4900
fax: 414 353 5231

Carpenter Technology Corp.
2809 Winter Street, N.E.
Minneapolis, MN 55413-2977
USA
phone: 612 331 8440
fax: 612 331 5662

Carpenter Technology Corp.
99 Clinton Road
P.O. Box 1267
West Caldwell, NJ 07006-1267
USA
phone: 201 227 5400
fax: 201 227 4098

Carpenter Technology Corp.
235 New York Drive
Fort Washington, PA 19034-2589
USA
phone: 215 646 7272
fax: 215 283 4599

Carpenter Technology Corp.
145 Metro Park
P.O. Box 23020
Rochester, NY 14692-3020
USA
phone: 716 475 9120
fax: 716 424 6972

Carpenter Technology Corp.
3942 Laclede Avenue
St. Louis, MO 63108-3273
USA
phone: 800 654 6543
fax: 708 681 1006

Carpenter Technology Corp.
2263 National Avenue
Unit B
Hayward, CA 94545-1715
USA
phone: 800 654 6543
fax: 213 587 8145

(continued)

Carpenter Technology Corp. (cont'd)

Carpenter Technology Corp.
1717 North Commerce Drive
South Bend, IN 46628-1594
USA

phone: 800 654 6543
fax: 313 588 6466

Castings USA

P.O. Box 7677
Pine Bluff, AR 71611
USA

phone: 501 534 0141

Cast iron

Central Engineering & Supply Co.

834 Main Ave.
Passaic, NJ 07055
USA

phone: 201 777 9000
fax: 201 777 8946

Tin and alloys

Ceradyne, Inc.

3169 Redhill Ave.
Costa Mesa, CA 92626
USA

phone: 714 549 0421
fax: 714 549 5787

Mechanical ceramics; Electrical ceramics; Carbides; Nitrides; Oxides

Cercast Group

An Affiliate of Howmet Corp.
3905 Industrial Blvd.
Montreal-North, Quebec H2H 2Z2
Canada

phone: 514 322 2371
fax: 514 322 1340

Copper and alloys

Cerro Copper Products Co.

P.O. Box 66800
St. Louis, MO 63166-6800
USA

Copper and alloys

Cerro Metal Products Co.

P.O. Box 388
Rt. 144 South
Bellefonte, PA 16823
USA

phone: 814 355 6357
fax: 814 355 6227

Tool steels; Copper and alloys; Lead and alloys; Zinc and alloys

Certified Alloy Products, Inc.

3245 Cherry Ave.
P.O. Box 90
Long Beach, CA 90801
USA

phone: 310 979 7101
fax: 310 427 8667

Stainless steels, cast; Nickel, alloys, superalloys; Refractory metals

CEZUS

Subsidiary of Pechiney Group
Tour Manhattan
Cedex 21
F92087 Paris La Defense
France

phone: 33 1 47 62 88 00
fax: 33 1 47 62 86 71

Titanium and alloys; Zirconium

Chace Precision Metals

Subsidiary of Technitrol, Inc.
1704 Barnes Street
Reidsville, NC 27323
USA

phone: 919 342 2381

Ferrous electronic materials; Manganese

Chambersburg Engineering Co.

150 Derbyshire St.
P.O. Box V
Chambersburg, PA 17201
USA

phone: 717 264 7151
fax: 717 267 2201

Cast iron

Champion Rivet Co.

A Div. of Hogen Industries
4010 E. 116th Street
Cleveland, OH 44105
USA

phone: 800 348 6014
fax: 216 241 7983

Ferrous welding, hard facing electrodes

Champion Steel Co.

P.O. Box 97
8247 Penniman Road
Orwell, OH 44076
USA

phone: 216 437 5161
fax: 216 437 5163

Tool steels

Chase Brass

Montpelier, OH
USA

phone: 419 485 3193

Copper and alloys

Chemalloy Co., Inc.

P.O. Box 350
996 RR Avenue
Bryn Mawr, PA 19010-0350
USA

phone: 215 527 3700
fax: 215 527 3878

Aluminum and alloys; Cadmium, chromium; Cobalt, alloys, superalloys; Magnesium and alloys; Nickel, alloys, superalloys; Titanium and alloys; Carbides

Chemfab Corp.

701 Daniel Webster Hwy.
PO Box 1137
Merrimack, NH 03054
USA

phone: 800 262 2990
fax: 603 424 9028

Fluoropolymers, chloropolymers

Chevron Chemical Company

Po Box 3766
Houston, TX 77253
USA

phone: 800 231 4638
fax: 713 754 4479

Polyolefins; Polystyrenes

China Metallurgical Import & Export Corp

46 Dongsixi Xidajie
Beijing
China

phone: 86 1 550197

Carbon steels, mill shapes; Carbon steels, sheet, strip; Carbon steels, plate; Alloy steels, mill shapes; Cast iron; Tool steels; Refractory ceramics; Carbon, graphite, diamond; Oxides

China Steel Corporation

Lin Hai Industrial District
P.O. Box 47-29
Hsiao Kang, Kaohsiung 81233
Taiwan, Republic of China

phone: 886 07 8021111
fax: 886 07 8022511

Carbon steels, mill shapes; Carbon steels, sheet, strip; Carbon steels, plate; Alloy steels, sheet, strip; Alloy steels, plate; Aluminum and alloys

Chris Craft Industrial Products, Inc.

407 County Line Road
Gary, IN 46403-2699
USA

phone: 219 762 3165
fax: 219 763 4477

Polyvinyl chlorides

OTHER LOCATION

Chris Craft Industrial Products, Inc.

450 W. 169th St.
South Holland, IL 60473
USA

Chromium Corp. of America

8701 Union Ave.
Cleveland, OH 44105
USA

phone: 216 271 4910
fax: 216 429 0637

Cadmium, chromium

Ciba Geigy Corporation

Composite Materials
5115 East LaPalma Ave.
Anaheim, CA 92807-2018
USA
phone: 800 854 1869
fax: 714 779 7183

Epoxies; Polyesters, thermoset; Carbon, graphite, diamond; Thermoplastic elastomers; Fibers

OTHER LOCATIONS

Ciba Geigy Corporation
Furane Aerospace Products
5121 San Fernando Rd. West
Los Angeles, CA 90039-1011
USA
phone: 818 247 6210
fax: 818 507 0167

Ciba Geigy Corporation
Formulated Systems Group
4917 Dawn Ave.
East Lansing, MI 48823
USA
phone: 517 351 5900

Brochier S.A.
z.1 Chartinieres
01120 Dagneux
France
phone: 33 7806 2066
fax: 33 7806 0292

Ciba Geigy Corporation
Resins Division
7 Skyline Drive
Hawthorne, NY 10532
USA
phone: 914 347 4700

Ciba Composites
Duxford
Cambridge CB2 4QD
England
phone: 44 0223 833141
fax: 44 0223 838808

Danutec Werkstoff GmbH
Industriestr. 1
A-4061 Pasching
Austria
phone: 43 7229 772 0
fax: 43 7229 772 299

(continued)

Ciba Geigy Corporation (cont'd)

Heath Tecna Aerospace Company
19819 84th Avenue South
P.O. Box 97004
Kent, WA 98064-9704
USA

phone: 206 872 7500
fax: 206 395 1237

Climax Specialty Metals

21803 Tungsten Rd.
Cleveland, OH 44117
USA

phone: 216 692 3990
fax: 216 692 0029

Alloy steels, cast; Stainless steels, cast; Cast iron; Cobalt, alloys, superalloys; Copper and alloys; Refractory metals

CMW Inc.

P.O. Box 2266
Indianapolis, IN 46206
USA

phone: 317 634 8884
fax: 317 638 2706

Alloy steels; Copper and alloys; Precious metals; Refractory metals; Carbides

OTHER LOCATION
CMW Inc.
70 S. Gray Street
Indianapolis, IN 46201
USA

Cogebi, Inc.

14 Faraday Drive
Dover, NH 03820
USA

phone: 603 749 6896
fax: 603 749 6958

Engineering ceramics

Cogne SpA

Via Paravera 16
I-11100 Aosta
Italy

phone: 39 0165 3021
fax: 39 0165 43833

Alloy steels

Cold Metal Products Company, Inc.

P.O. Box 6078
45 S. Montgomery Avenue
Youngstown, OH 44501
USA

phone: 216 746 6311
fax: 216 742 1200

Carbon steels, sheet, strip; Alloy steels, sheet, strip

Colonial Metals Co.

P.O. Box 311
217 Linden St.
Columbia, PA 17512
USA

phone: 717 684 2311
fax: 717 684 9555

Copper and alloys

Colorite Plastics

101 Railroad Ave.
Ridgefield, NJ 07657
USA

phone: 201 941 2900
fax: 201 941 0308

Polyvinyl chlorides

Columbus McKinnon Corp.

Product Service
P.O. Box 1294
Abingdon, VA 24210
USA

phone: 800 634 4647
fax: 703 628 8653

Alloy steels P.O. Box 1294

Comalco Ltd.

GPO box 2773Y
Melbourne, Victoria 3001
Australia

phone: 61 03 658 8300
fax: 61 03 658 3903

Aluminum and alloys

OTHER LOCATION
Comalco Ltd.
55 Collins St.
Melbourne, Victoria 3001
Australia

fax: 61 658 3707

Comalloy International Corp.

481 Allied Dr.
Nashville, TN 37211
USA

phone: 615 333 3453
fax: 615 834 9941

Polyamides; Polycarbonates; Polyesters, thermoplastic; Polyolefins; Polystyrenes

Cominco Ltd.

Toronto Div.
120 Adelaide St. W.
Suite 1500
Toronto, Ontario M5H 1T1
Canada

phone: 416 869 1850
fax: 416 862 8553

Cadmium, chromium; Lead and alloys; Precious metals; Zinc and alloys

(continued)

Cominco Ltd. (cont'd)

OTHER LOCATIONS

Cominco Ltd.
Toronto Div.
120 Adelaide St. W.
Suite 1500
Toronto, Ontario M5H 1T1
Canada
phone: 416 869 1850
fax: 416 862 8553

Cominco Ltd.
Box 1000
Trail, British Columbi VIR 418
Canada
phone: 604 364 4222

Conap, Inc.

1405 Buffalo St.
Olean, NY 14760-1139
USA

phone: 716 372 9650
fax: 716 372 1594

Epoxies; Polyurethanes

Concept Polymer Technologies, Inc.

8625 Bryan Dairy Rd.
Largo, FL 34647
USA

phone: 800 541 6880
fax: 813 399 5505

Thermoplastic elastomers

Continental Foundry & Machine Co.

10062 Hwy 78
Olive Branch, MS 38654
USA

phone: 601 895 5551
fax: 601 895 5554

Alloy steels; Cast iron

Cook Composites and Polymers

217 Freeman Drive
PO Box 996
Port Washington, WI 53074
USA

phone: 414 284 5541
fax: 414 284 7517

Acrylics; Polyesters, thermoset

Cook, George H. & Co. Ltd.

Uniformity Steel Works
P.O. Box 22
Corporation St.
Sheffield, S Yorks S3 8RE
England

phone: 44 0742 737603
fax: 44 0742 755935

Tool steels

Coors Ceramics Company

600 9th St.
Golden, CO 80401
USA

phone: 303 278 4000
fax: 303 277 4502

Electrical ceramics; Refractory ceramics; Engineering ceramics; High alumina and "cermets"

Copper Range Co.

P.O. Box 100
1 Wilcox Road
White Pine, MI 49971-0100
USA

phone: 906 885 5111
fax: 906 885 5105

Copper and alloys

Coshocton Stainless Div.

Armco Inc.
P.O. Box 548
17400 St.Rt. #16
Cochocton, OH 43812
USA

phone: 614 829 2341
fax: 614 829 2793

Stainless steels, mill shapes; Stainless steels, sheet, strip

Cosmic Plastics, Inc.

27939 Beale Court
Valencia, CA 91355
USA

phone: 800 423 5613
fax: 805 257 3345

Epoxies; Polyesters, thermoset

Cotronics Corp.

3379 Shore Parkway
Brooklyn, NY 11235
USA

phone: 718 646 7996
fax: 718 646 3028

Epoxies; High alumina and "cermets"; Oxides

Courtaulds Aerospace Inc.

P.O. Box 11
Foleshill Rd.
Coventry CV6 5AB
England

phone: 44 203 861861
fax: 44 203 861800

Fluoropolymers, chloropolymers; Polyamides;
Polyesters, thermoplastic; Glass, quartz; Fibers

(continued)

Courtaulds Aerospace Inc. (cont'd)

OTHER LOCATIONS

Courtaulds Aerospace Inc.
1608 4th Street
Berkley, CA 94710
USA
phone: 510 526 1525
fax: 510 525 5669

Courtaulds Aerospace Inc.
Hazelhurst Works
Lower Hollin Bank Street
Blackburn, Lancashire BB2 4UB
England
phone: 44 0254 680237
fax: 44 0254 682473

Courtaulds Aerospace Inc.
Electrical Insulation
Chadwick Road
Astmoor Industrial Estate
Runcorn, Cheshire WA7 1PW
England
phone: 44 0928 580508
fax: 44 0928 580100

Courtaulds Structural Composites

6 Shields Dr.
Bennington, VT 05201
USA
phone: 802 442 9964
fax: 802 447 3642

Polyolefins

OTHER LOCATION

Courtaulds Structures Composites S.A.
53 rue Daniele-Casanova
76410 Tourville la Riviere
France
phone: 33 35 77 58 97
fax: 33 35 78 53 81

Crane Valves

Jenkin Bros.
104 N. Chicago St.
Joliet, IL 60431
USA

phone: 815 727 2600
fax: 815 727 4054

Cast iron

Crane, Torry S., Co.

P.O. Box 374
492 Summer Street
Plantsville, CT 06479
USA

phone: 203 628 4778

Stainless steels; Stainless steels, cast; Cast iron; Copper and alloys; Nickel, alloys, superalloys

Criterion Metals Inc.

Attleboro Industrial Park
35 Extension Street
Attleboro, MA 02703-4642
USA

phone: 508 761 6396

Copper and alloys; Nickel, alloys, superalloys

Crucible Materials Corp.

Box 977
State Fair Blvd.
Syracuse, NY 13201-0977
USA

phone: 315 487 4111

Alloy steels, mill shapes; Stainless steels

Cryovac

Div. of W. R. Grace & Co.-Conn.
PO Box 464
Duncan, SC 29334-0464
USA

phone: 803 433 2000

Polyolefins

Custom Coating and Laminating Corp.

717 Plantation St.
Worcester, MA 01605
USA

phone: 508 852 3072
fax: 508 852 3759

Polyesters, thermoplastic; Polystyrenes

Custom Resins

Div. of Bemis Co., Inc.
PO Box 46
Henderson, KY 42420
USA

phone: 800 626 7050
fax: 502 827 8509

Polyamides

Cyprus Foote Mineral Co.

Suite 301
301 Lindenwood Drive
Malvern, PA 19355-1740
USA

phone: 215 889 9605
fax: 215 889 9612

Lithium; Ceramics

CYRO Industries

100 Valley Road
PO Box 950
Mt.Arlington, NJ 07856
USA

phone: 201 770 3000
fax: 201 770 6117

Polystyrenes

Cytec Industries

a business unit of American Cyanamid
PO Box 148
12600 Eckel Rd.
Perrysburg, OH 43551
USA

phone: 419 874 7941
fax: 419 874 0951

Polyesters, thermoset

Cytemp Specialty Steel

a Division of Armco Inc.
P.O. Box 247
701 East Spring Street
Titusville, PA 16354
USA

phone: 814 827 3641
fax: 814 827 3951

Stainless steels

D-M-E Company

A Fairchild Industries Co.
29111 Stephenson Highway
Madison Heights, MI 48071
USA

phone: 313 398 6000
fax: 313 544 5705

Alloy steels; Stainless steels

(continued)

D-M-E Company (cont'd)

OTHER LOCATIONS

All-Ways Engineering Services Ltd.
P.O. Box 19.181
7 Wingate Street
Auckland 7
New Zealand
phone: 64 9 8287035
fax: 64 9 8281118

Amalgamated Diemould D-M-E Pty. Ltd.
10 Warren Avenue
Bankstown, N.S.W. 22200
Australia
phone: 61 2 7081173
fax: 61 2 7086964

Biasi Edelstahl AG
Postfach 184
Bahnofstrasse 182
CH 8622 Wetzikon
Switzerland
phone: 41 1 9304343
fax: 41 1 9301315

Comat D-M-E S.P.A.
Via Desiderio 24
20131 Milano
Italy
phone: 39 2 2665494
fax: 39 2 2665142

D-M-E (H.K.) Ltd.
Unit 4, G/F Po Lung Centre
11 Wang Chiu Rd.
Kowloon Bay
Kowloon
Hong Kong
phone: 852 7951035
fax: 852 7547698

D-M-E Belgium
D-M-E Europe (Headquarters)
Industriepark, Noord
B-2800 Mechelen
Belgium
phone: 32 15 215011
fax: 32 15 218235

D-M-E Engineering Pte. Ltd.
No. 10 Wan Shih Road
Jurong Industrial Estate
Singapore 2262
Singapore
phone: 65 2640811
fax: 65 2659497

D-M-E Europe U.K.
Halifax Road
Cresex Industrial Estate
High Wycombe
Bucks HP12 3TN
England
phone: 44 494 534074
fax: 44 494 447374

(continued)

D-M-E Company (cont'd)

D-M-E France, S.A.R.L.
Parc D Activities De La Danne
BP 100
Eragny-Sur-Oise
95613 Cergy Pontoise Cedex
France
phone: 33 13 4483160
fax: 33 13 4488922

D-M-E Iberica S.A.
Industrialdea No. 3
20180 Oyarzun (Guipuzocoa)
Spain
phone: 34 43 492250
fax: 34 43 490005

D-M-E Normalien
7106 Nuenstadt
AM Kocher
Germany
phone: 49 71 39920
fax: 49 71 399240

D-M-E of Canada Ltd.
6210 Northwest Drive
Mississauga, Ontario
Canada
phone: 416 677 6370
fax: 416 677 5280

FKL Taiwan Steel Co.
No. 72, Wuu-Chueng 7th Rd.
Wuu-Guu Industrial Park
Taipei
Taiwan R.O.C.
phone: 886 2 2996787
fax: 886 2 2997608

Hestico (Pty.) Ltd.
P.O. Box 1985
Bramley 2018, Johannesburg
South Africa
phone: 27 11 7865270
fax: 37 11 8851693

Japan D-M-E Corp.
Harumi Park Bldg., 4th Floor
2-22, Harumi 3-Chome
Chuo-ku
Tokyo
Japan
phone: 81 3 5332281
fax: 81 3 5317849

Leong Bee & Soo Bee SDN, BHD.
1-3 MacCallum St., Ghaut
P.O. Box 23
Penang
Malaysia
phone: 60 4 624226
fax: 60 4 622661

Marshall Trading Co. Ltd.
Factopia Suite #305
196 Anyang-Dong
Anyang-City, Kyungki-Do
Korea
phone: 82 343 672500
fax: 82 343 672511

MESCO
P.O. Box 1688 MCPO.
1256 Makati, Metro Manilla
Philippines
phone: 63 2 6734301
fax: 63 2 6733090

(continued)

D-M-E Company (cont'd)

Partes Para Moldes D-M-E, S.A. De C.V.
Avenida 1 Lote 10-A
Parque Industrial Cartagena
Tultitlan
Estado De Mexico
Mexico
phone: 52 5 8720332
fax: 52 5 8724145

Polimold Industria De Moldes Ltda.
Rua Vietra De Morais
309 Campo Belo
04617 Sao Paulo
Brazil
phone: 55 11 2410955
fax: 55 11 5427800

Productos Humar, C.A.
APDO 60900
Caracas 1060A
Venezuela
phone: 58 2 325040
fax: 58 2 2620402(2700)

Samick Chemical Co. PVT. Ltd.
129, Pragati Industrial Estate
316, N.M. Johsi Marg
Bombay 400 011
India
phone: 91 22 3080763/64
fax: 91 22 3080206

Soteime LTDA.
Rua Sociedade Farmaceutica
N-20, 4B
Lisbon
Portugal
phone: 351 1 574545
fax: 351 1 563694

Daido Steel Co. Ltd.

Kogin Bldg., 11-18 Nishiki 1-Ch
Naka-ku
Nagoya 460
Japan

phone: 81 052 201 5111
fax: 81 052 211 1664

Carbon steels, mill shapes; Carbon steels, sheet, strip; Alloy steels, mill shapes; Alloy steels, sheet, strip; Aluminum and alloys

Dart Polymers, Inc.

60 E. Main St.
Reola, PA 17540
USA

phone: 800 723 1036

Polystyrenes

Deerfield Urethane, Inc.

1080 Grenoble Road
Ivyland, PA 18974
USA

phone: 215 953 8580
fax: 215 953 8582

Polyurethanes

Degussa AG

Geschaftsbereich Dental
Zweigniederlasoung Wolfgang
DE-VL Bldg. 1151
Rodenbacher Chaussee 4
D-63457 Hanau/Wolfgang
Germany

phone: 49 69 218 01
fax: 49 69 218 3218

Precious metals

Degussa AG (cont'd)

OTHER LOCATION

Degussa Corp.
Dental Dept.
3900 S. Clinton Ave.
South Plainfield, NJ 07080
USA

phone: 800 221 0168
fax: 908 754 6440

Degussa Corporation

Chemicals Division
65 Challenger Road
Ridgefield Park, NJ 07660
USA

phone: 201 641 6100
fax: 201 807 3183

Acrylics

Deloro Stellite Inc.

Div. of Thermadyne Industries Inc.
101 S. Hanley Road
Suite 300, 3rd Floor
St. Louis, MO 63105
USA

phone: 314 862 2666
fax: 314 721 4822

Stainless steels; Stainless steels, cast; Cast iron; Cobalt, alloys, superalloys; Nickel, alloys, superalloys

(continued)

Deloro Stellite Inc. (cont'd)

OTHER LOCATIONS

Deloro Stellite GmbH
Postfach 520
Zur Berflege 53
Germany
phone: 49 0 261 80880
fax: 49 0 793 823814

Deloro Stellite Ltd.
Stratton St. Margaret
Swindon
Wiltshire SN3 4QA
England
phone: 44 793 822451
fax: 44 793 823814

Deloro Stellite Inc.
471 Dundas Street East
Bellevile, Ontario K8N 5C4
Canada
phone: 613 968 3481
fax: 613 966 8269

Stellite Coatings Inc.
1201 Eisenhower Drive
Goshen, IN 46526
USA
phone: 219 534 2585
fax: 219 534 3417

Delsteel Inc.

P.O. Box 1268
Wilmington, DE 19899
USA

phone: 302 764 2200
fax: 302 764 7643

Tool steels

Delta (Manganese Bronze) Ltd.

Handford Works
P.O. Box 22
Hadleigh Road
Ipswich, Suffolk IP2 0EG
England

phone: 44 0473 252127
fax: 44 0473 218 229

Copper and alloys

Delta Enfield Metals Ltd.

Millmarsh Lane
Brimsdown
Enfield, Middlesex EN3 7QB
England

phone: 44 081 804 1255
fax: 44 081 804 5206

Copper and alloys

Delta Extruded Metals Company Ltd.

Greets Green Road
West Bromwich
West Midlands B70 9ER
England

phone: 44 021 500 6188
fax: 44 021 553 6505

Copper and alloys

Denver Thomas, Inc.

P.O. Box 96
Birmingham, AL 35201
USA

phone: 205 599 6600

Cast iron

Deutsche Nickel AG

Postfach 1840
Resenweg 15
5840 Schwerte 1
Germany

phone: 49 023 04 108 0
fax: 49 023 04 86164/86101

Alloy steels, mill shapes; Alloy steels, sheet, strip; Stainless steels, mill shapes; Stainless steels, sheet, strip; Copper and alloys; Nickel, alloys, superalloys

Dewramet Ltd.

York house
Mossland Road
Glasgow G52 4UD
Scotland

phone: 44 041 882 9001
fax: 44 041 883 2047

Iron-base superalloys; Cobalt, alloys, superalloys; Nickel, alloys, superalloys

Diamond Metal Alloys

4051 Hollister
Houston, TX 77080
USA

phone: 713 460 5332

Carbides

Diamond Polymers Inc.

1353 Exeter Road
Akron, OH 44306
USA

phone: 216 773 2700
fax: 216 773 2799

Polystyrenes

Diederichs, Karl Stahl-Walz und Hammerwe

Dirostahl Division
Postfaach 120120
D-5630 Remscheid 11
Germany

phone: 49 0 21 91 593 0
fax: 49 0 21 91 593 165

Carbon steels, mill shapes; Alloy steels, mill shapes

Division Lead Limited Partnership

7742 W. 61st Place
Summit, IL 60501
USA

phone: 800 457 5323
fax: 708 458 4642

Lead and alloys

DoAll Company

254 N. Laurel Ave.
Des Plaines, IL 60016
USA

phone: 708 824 1122
fax: 708 699 7524

Tool steels; Carbides

Dofasco Inc.

P.O. Box 2460
1330 Burlington Street
Hamilton, ONT L8N 3J5
Canada

phone: 416 544 3761
fax: 416 548 4293

Carbon steels, mill shapes; Alloy steels, mill shapes

Dorrenberg Edelstahl GmbH

Postfach 2164
Hammerweg 7
D-5250 Engelskirchen
Germany

phone: 49 0 22 63 50 21

Alloy steels; Stainless steels; Tool steels

Dow Chemical Co.

Plastics Group
P.O. Box 1206
Midland, MI 48641-1206
USA

phone: 800 441 4369

Magnesium and alloys; Polycarbonates; Polyurethanes; Polystyrenes

OTHER LOCATION

Dow Chemical Co.
Lake Jackson Research Ctr.
Bldg. 212
Freeport, TX 77541
USA

phone: 409 299 2365
fax: 409 299 2383

Dow Corning Corp.

Epoxy Resins & Intermediate Pr
PO Box 994
Midland, MI 48686-0994
USA

phone: 517 496 6000
fax: 517 496 4586

Polyolefins; Polyurethanes; Polystyrenes; Silicones

DSM Engineering Plastics

PO Box 3333
2267 West Mill Road
Evansville, IN 47732-3333
USA

phone: 800 333 4237
fax: 812 424 0892

Acetals; Polyamides; Polyolefins; Polystyrenes

OTHER LOCATIONS

DSM Engineering Plastics
26877 Northwestern Highway
Suite 410
Southfield, MI 48034
USA
phone: 313 827 6600
fax: 313 827 6605

DSM Engineering Plastics
P.O. Box 15051
501 Crescent Avenue
Reading, PA 19612
USA
phone: 800 366 0300
fax: 215 320 6930

DSM Engineering Plastics
233 Arvin Avenue
Stoney Creek, Ontario L8E 2L9
Canada
phone: 416 662 1866
fax: 416 662 3493

DSM Engineering Plastics
14540 Alondra Blvd.
La Mirada, CA 90638
USA
phone: 800 824 4381
fax: 714 670 7888

DSM Engineering Plastics
P.O. Box 43
6130 AA Sittard
Netherlands
phone: 31 46 77 01 23
fax: 31 46 77 04 00

DSM Japan K.K.
Shin Kokusai Bldg. 7F
4-1 Marunouchi
Tokyo 100
Japan
phone: 81 03 3217 8941
fax: 81 03 3201 5074

DSM Engineering Plastics
152 Beach Road
Gateway East
Singapore 0718
Singapore
phone: 65 299 6080
fax: 65 299 5848

Dupont, E.I., de Nemours & Co.

Metal Cladding Section
220 Gale Lane
Suite 5
Kennett Square, PA 19348
USA

phone: 800 441 7515
fax: 215 444 2020

Stainless steels; Nickel, alloys, superalloys; Refractory metals; Titanium and alloys

OTHER LOCATION
Dupont Company
Films Enterprise
Kirk Mill Bldg.
Barley Mill Plaza
Wilmington, DE 19880
USA

phone: 800 441 9494
fax: 302 999 4358

Dural Aluminum Composites Corp.

10505 Roselle St.
San Diego, CA 92121
USA

phone: 619 587 1415
fax: 619 587 0920

Aluminum and alloys

Duraloy Blaw-Knox Inc.

Div. of Blaw Knox Corp.
P.O. Box 81
120 Bridge St.
Scottdale, PA 15683
USA

phone: 412 887 5100
fax: 412 887 5224

Alloy steels, cast; Stainless steels, cast; Stainless steels, mill shapes; Cast iron

Duriron Co. Inc.

425 N. Findlay St.
P.O. Box 1145
Dayton, OH 45401
USA

phone: 513 226 4523

Alloy steels; Iron-base superalloys

DWA Composite Specialties, Inc.

21119 Superior St.
Chatsworth, CA 91311
USA

phone: 818 998 1504
fax: 818 998 0343

Aluminum and alloys; Copper and alloys; Magnesium and alloys

Dyn-Metal Ltd.

25-29 Chase Road
North Acton
London NW10 6TA
England

phone: 44 81 961 0656
fax: 44 81 961 8820

Copper and alloys

(continued)

Dyn-Metal Ltd. (cont'd)

OTHER LOCATIONS

Pan-Metallgesellschaft Baumgaertner & Co K.G.
Postfach 102436
D-6800 Mannheim
Germany
phone: 49 621423030
fax: 49 6214230333

Societe Metallurgique
De Bronze Pan
1 Rue Poincare
57206 Sarreguemines
France
phone: 33 87982076
fax: 33 87951814

Eagle Alloy Inc.

5142 Evanston Ave.
Muskegon, MI 49442
USA

phone: 616 788 2351
fax: 616 788 5515

Carbon steels, cast; Alloy steels, cast; Stainless steels, cast

Eastern Stainless Corp.

Division of Armco Inc.
P.O. Box 1975
Baltimore, MD 21203
USA

phone: 800 543 1234
fax: 410 522 6434

Stainless steels, mill shapes; Stainless steels, sheet, strip; Stainless steels, plate

Eastman Chemical Company

Performance Plastics Business
P.O. Box 1969
137 Regional Park Dr.
Kingsport, TN 37662
USA

phone: 800 327 8626
fax: 615 229 8595

Polyesters, thermoplastic; Polyolefins

Elf Atochem North America Inc.

2000 market Street
Philadelphia, PA 19103-3222
USA

phone: 800 628 3391
fax: 215 419 7400

Polyimides

OTHER LOCATION

Elf Atochem S.A.
4, cours Michelet
La Defense 10 - Cedex 42
92091 Paris La Defense
France

phone: 33 1 49 00 80 80
fax: 33 1 49 00 80 44

Empire Steel Castings Inc.

P.O. Box 139
Reading, PA 19603
USA

phone: 215 921 8101
fax: 215 921 9838

Stainless steels, cast; Alloy steels, cast; Nickel, alloys, superalloys; Carbon steels, cast

Empire-Detroit Steel Div.

Division of Armco Inc.
913 Bowman Street
Mansfield, OH 44903
USA

phone: 800 672 5010
fax: 419 755 3441

Carbon steels, sheet, strip

OTHER LOCATIONS

Empire-Detroit Steel
Division of Armco Inc.
303 Oxford Street
Dover, OH 44622
USA
phone: 216 343 3371
fax: 216 364 3472

Empire-Detroit Steel
Bowman Metal Deck Products
913 Bowman Street
Mansfield, OH 44903
USA
phone: 419 755 3258
fax: 419 755 3430

EMS-American Grilon, Inc.

PO Box 1717
Sumter, SC 29151-1717
USA

phone: 803 481 9173
fax: 803 481 3820

Polyamides

Engelhard Corporation

Engineered Materials Div.
700 Blair Road
Carteret, NJ 07008
USA

phone: 908 205 7451
fax: 908 205 7453

Cadmium, chromium; Copper and alloys; Precious metals; Refractory metals

OTHER LOCATIONS

CMP Engelhard
4 rue Beaubourg
75004 Paris
France
phone: 33 14 461 1000
fax: 33 14 670 7600

Engelhard (Hong Kong) Ltd.
Eldex Industrial Bldg.
6th Floor, Block B
21 Mau Tau Wei Road
Hunghom, Kowloon
Hong Kong
phone: 852 3650301
fax: 852 7656406

Engelhard (Singapore) Pte. Ltd.
8 Kaki Bukit Road 2#
02-35/26 Ruby Warehouse Complex
Singapore 1441
Singapore
phone: 65 743 6918
fax: 65 744 8424

Engelhard Australia Pty. Ltd.
10-12 Prospect St.
Box Hill 3128, Victoria
Australia
phone: 61 3 899 6330
fax: 61 3 899 6360

(continued)

Engelhard Corporation (cont'd)

Engelhard Corporation
700 Blair Road
Cateret, NJ 07008
USA
phone: 908 205 7451
fax: 908 205 7453

Engelhard SpA
Via di Salone
245 Settecamini
00131 Rome
Italy
phone: 39 8 4199 2124
fax: 39 8 4199 2275

Engelhard GmbH
Lise Meltner Srasse 7
6072 Dreleich
Germany
phone: 49 6103 35072
fax: 49 6103 34787

Engelhard West, Inc.
4025 Clipper Court
Fremont, CA 94538
USA
phone: 510 490 2150
fax: 510 490 0851

Engelhard Limited
Davis Road
Chessington, Surrey KT9 1XQ
England
phone: 44 81 397 5292
fax: 44 81 397 7230

Engelhard of Canada Ltd.
100 Engelhard Drive
P.O. Box 340
Aurora, Ontario L4G 3N1
Canada
phone: 416 727 3191
fax: 416 727 1078

N.E. Chemoat Corp.
P.O. Box 94
World Trade Center Bldg.
4-1 Hamamatsucho, 2-Chome
Minato-Ku, Tokyo 105
Japan
phone: 81 03 3435 5492
fax: 81 03 3435 5484

Enichem Tecnoresine SpA

1211 Avenue of the Americas
9th Floor
New York, NY 10036
USA

phone: 212 382 6300
fax: 212 382 6520

Polyamides; Polycarbonates; Polystyrenes

OTHER LOCATION
Enichem Tecnoresine SpA
Via Medici del Vascello, 40
20138 Milano
Italy

phone: 39 02 5201
fax: 39 02 5203 9590

Enpar Sonderwerkstoffe GmbH

Postfach 1144
Overather StraBe 37
D-5250 Engleskirchen
Germany

phone: 49 02263/30 71
fax: 49 02263 20685

Stainless steels; Nickel, alloys, superalloys; Titanium and alloys; Zirconium

Epic Resins

Div. of General Fiberglass Sup
1415 Ellis St.
Waukesha, WI 53186
USA

phone: 414 521 2255
fax: 414 521 1320

Epoxies

Erasteel Champagnole

a subsidiary of Erasteel
23 Rue Clemenceau
BP 104
F39300 Champagnole
France

phone: 33 084 52 64 44
fax: 33 084 52 61 25

Tool steels

OTHER LOCATIONS

Erasteel (U.K.) Ltd.
371 Coleford Road Darnall
Sheffield S9 SNF
England
phone: 44 742 610 410
fax: 44 742 617 167

Erasteel Commentry
BP 1
1, place Martenot
03600 Commentry
France
phone: 33 70 28 78 00
fax: 33 70 28 78 93

Erasteel GmbH
Postfach 430
Westring 9
4010 Hilden
Germany
phone: 49 2103 49 05 00
fax: 49 2103 49 05 21

Erasteel Inc.
100 New Dutch Lane
Fairfield, NJ 07006
USA
phone: 201 237 6444
fax: 201 227 4115

Erasteel Italiana S.R.L.
Via XXV Aprile 150
100 42 Nichelino, Torino
Italy
phone: 39 11 62 70 139
fax: 39 11 62 11 32

Erasteel Kloster AB
Box 100
81060 Soderfors
Sweden
phone: 46 293 173 50
fax: 46 293 307 70

Erasteel Kloster AB
Box 102
770 70 Langshyttan
Sweden
phone: 46 225 630 00
fax: 46 225 616 49

Erasteel Kloster AB
Box 101
77602 Vikmanshyttan
Sweden
phone: 46 225 306 60
fax: 46 225 304 60

(continued)

Erasteel Champagnole (cont'd)

Erasteel Kloster AB
Soderfors Powder
Box 100
810 60 Soderfors
Sweden

phone: 46 293 173 50
fax: 46 293 307 70

Erbsloh, J. & A., GmbH & Co.

Press und Veredlungswerk
Siebeneicker Strasse 235
P.O. Box 150160
D-2053 Velbert 15, (Neviges)
Germany

phone: 49 02053 491049
fax: 49 02053 491281

Aluminum and alloys

ERG Materials & Aerospace Corp.

900 Stanford Ave.
Oakland, CA 94608
USA

phone: 510 658 9785
fax: 510 658 7428

Aluminum and alloys; Carbides; Carbon, graphite, diamond

Ergst, Stahlwerke

Kommanditgesellschaft
Letmather Str. 69
5840 Schwerte 1
Germany

phone: 49 0 2304 790
fax: 49 0 2304 793 90

Carbon steels, mill shapes; Carbon steels, sheet, strip; Stainless steels; Stainless steels, mill shapes; Stainless steels, sheet, strip

ESAB Group

Subsidiary of The ESAB Group
801 Wilson Ave.
P.O. Box 517
Hanover, PA 17331
USA

phone: 800 933 7070
fax: 717 637 9859

Ferrous welding, hard facing electrodes

OTHER LOCATIONS

Welding Consumables Inc.
Subsidiary of The ESAB Group
801 Wilson Ave.
P.O. Box 517
Hanover, PA 17331
USA
phone: 800 933 7070
fax: 717 637 9859

All-State Welding Products, Inc.
Subsidiary of The ESAB Group
5112 Allendale Lane
P.O. Box 600
Taneytown, MD 21787-0600
USA
phone: 410 756 4330
fax: 410 756 4333

ESCO Corp.

P.O. Box 10123
2141 NW 25th Avenue
Portland, OR 97210
USA

phone: 503 228 2141
fax: 503 778 6754

Alloy steels, cast; Stainless steels, cast; Magnesium and alloys

Europa Metalli-LMI SpA

50121 Firenze
Borgo Pinti 97/99
Italy

phone: 39 055 497 41
fax: 39 055 234 2214

Copper and alloys

Eutectic Corp.

Member of Eutectic + Castolin Companies
40-40 172nd St.
Flushing, NY 11358-9981
USA

phone: 800 221 1433
fax: 718 358 4628

Ferrous welding, hard facing electrodes; Aluminum and alloys; Copper and alloys; Nickel, alloys, superalloys; Precious metals

(continued)

Eutectic Corp. (cont'd)

OTHER LOCATIONS

Castolin Espana SA
Saturnino Calleja 7
28002 Madrid
Spain

Castolin GmbH
Gutenbergstrasse 10
D-6239 Kriftel
Germany

Castolin SA
Case Postale 360
CH-1001 Lausanne
Switzerland

Centre for TeroTechnology of Australia Pty., Ltd.
P.O. Box 249
9 Mavis Street
Bankstown, NSW 2200
Australia

EIM Industrias Metalugicas, Ltda.
Avenida do Rio Bonito 1.109
Caixa Postal 6707
Socorro, Santo Amaro
04776 Sao Paulo
Brazil

EM Industrias Metalurgicas SA
KM 36.5 Autopista Mexico-Quertaro
54730 Cuautitlan Izcalli
Estado de Mexico
Mexico

Eutectic Canada Inc.
4200 Route Trans Canada Highway
Pointe Claire, Quebec H9R 1B6
Canada

Eutectic Company Ltd.
Burnt Meadow Road
North Moons Moat
Redditch, Worcester B98 9NZ
U.K.

Eutectic of Japan, Ltd.
1224 Ogura
Saiwai-ku
Kawasaki 211
Japan

S.A. TeroTechnologie N.V.
224 Blvd. de l'Humanite
B-1190 Bruxelles
Belgium

Societe Francaise de TeroTechnologie
Avenue du Quebec
B.P. 325
Z.A. Courtaboef-Villebon
F-91958 Les Ulis Cedex
France

Faitout Iron & Steel Co.

182 Frelinghuysen Ave.
Newark, NJ 07114
USA

phone: 201 242 1000
fax: 201 242 2246

Alloy steels, mill shapes; Tool steels

Falk Corporation, The

Subs. of Sundstrand Corp.
Box 492
Milwaukee, WI 53201
USA

phone: 414 342 3131
fax: 414 937 4359

Alloy steels, cast

Fansteel VR/Wesson

c/o Fansteel Metals
Number One Tantalum Place
North Chicago, IL 60064
USA

phone: 708 689 4900

Nickel, alloys, superalloys; High alumina and "cermets"; Carbides

Fansteel Wellman Dynamics

U.S. Rt. 34 East
P.O. Box 147
Creston, IA 50801
USA

phone: 515 782 8521
fax: 515 782 4844

Aluminum and alloys; Magnesium and alloys

Federal-Mogul Corp.

Engine & Transmission Product Group
P.O. Box 1966
N-3
Detroit, MI 48235
USA

phone: 313 354 7700
fax: 313 354 9936

Alloy steels, cast; Aluminum and alloys; Copper and alloys; Lead and alloys; Tin and alloys; Zinc and alloys

Ferro Corporation

Filled and Reinforced Plastics
5001 O'Hara Drive
Evansville, IN 47711
USA

phone: 812 423 5218
fax: 812 423 1864

Epoxies; Polyesters, thermoset; Polyimides; Polyolefins

Ferry-Capitain, Usines De Bussy

Boite Postale 33-52
Joinville
France

Stainless steels, cast

Fiber Materials Inc.

Biddeford Industrial Park
5 Morin Street
Biddeford, ME 04005
USA

phone: 207 282 5911
fax: 207 282 7529

Epoxies; Polyimides

Fina Oil & Chemical Co.

P.O. Box 2159
Dallas, TX 75221
USA

phone: 214 750 2400
fax: 214 750 2392

Polyolefins; Polystyrenes

(continued)

Fina Oil & Chemical Co. (cont'd)

OTHER LOCATIONS

Fina Oil & Chemical Co.
22 Waterville Road
Avon, CT 06001
USA
phone: 203 674 1443

Fina Oil & Chemical Co.
8350 N. Central Expwy.
Suite 1600
Dallas, TX 75221
USA
phone: 800 362 0996

Fina Oil & Chemical Co.
2970 Parrott Avenue NW
Atlanta, GA 30318
USA
phone: 404 799 5074

Fina Oil & Chemical Co.
La Porte Research & Technical
Highway 134 at Miller Cut-off
P.O. Box 1200
Deer Park, TX 77536
USA
phone: 713 884 0560

Fina Oil & Chemical Co.
1450 East American Lane
Zurich Tower II
Suite 1325
Schaumburg, IL 60173
USA
phone: 708 240 1580

Finkl, A., & Sons Co.

2011 N. Southport Ave.
Chicago, IL 60614
USA

phone: 312 975 2510
fax: 312 348 5347

Tool steels

Firestone Synthetic Rubber & Latex

Div. of Firestone Tire & Rubber
381 W. Wilbeth Rd.
PO Box 26611
Akron, OH 44319-0006
USA

phone: 800 282 0222
fax: 216 379 7875

Polystyrenes

Firth-Vickers Special Steels Ltd.

Staybrite Works
Weedon St.
P.O. Box 131
Sheffield S9 2FU
England

phone: 44 0742 449955
fax: 44 0742 445390

Stainless steels, mill shapes; Tool steels; Nickel, alloys, superalloys

(continued)

Fisher Scientific Co.

711 Forbes Ave.
Pittsburgh, PA 15219
USA

phone: 412 562 8300
fax: 412 562 8313

Stainless steels, mill shapes; Zinc and alloys

OTHER LOCATIONS

Fisher Scientific Co.
2775 Pacific Drive
Norcross, GA 30091
USA
phone: 404 449 5050

Fisher Scientific Co.
900 Stewart Ave.
Plano, TX 75074
USA
phone: 214 578 0666

Fisher Scientific Co.
5637 Galleria Drive
Suite 101
Baton Rouge, LA 70816
USA
phone: 504 295 0401

Fisher Scientific Co.
5475 Peoria Street
Building 2, Unit J
Denver, CO 80239
USA
phone: 303 371 0888

Fisher Scientific Co.
1600 W. Glenlake Ave.
Itasca, IL 60143
USA
phone: 708 773 3050

Fisher Scientific Co.
10700 Rockley Road
Houston, TX 77099
USA
phone: 713 495 6060

Fisher Scientific Co.
9403 Kenwood Rd.
Suite C-208
Cincinnati, OH 45242
USA
phone: 513 793 5100

Fisher Scientific Co.
2761 Walnut Ave.
Tustin, CA 92681
USA
phone: 714 669 4600

(continued)

Fisher Scientific Co. (cont'd)

Fisher Scientific Co.
52 Fadem Road
Springfield, NJ 07081
USA
phone: 201 467 6400

Fisher Scientific Co.
7445 Exchange Drive
Orlando, FL 32809
USA
phone: 407 857 3600

Fisher Scientific Co.
50 Valley Stream Parkway
Great Valley Corporate Ctr.
Malvern, PA 19355
USA
phone: 215 640 7940

Fisher Scientific Co.
585 Alpha Drive
Pittsburgh, PA 15238
USA
phone: 412 963 3300

Fisher Scientific Co.
Carretera No. 1 KM 56.4
Barrio Montellano
P.O. Box 1760
Cayey, Puerto Rico 00736
USA
phone: 809 738 4231

Fisher Scientific Co.
1241 Ambassador Blvd.
St. Louis, MO 63132
USA
phone: 314 991 2400

Fisher Scientific Co.
2170 Martin Avenue
Santa Clara, CA 95050
USA
phone: 408 727 0660

Fisher Scientific Co.
112 ch. Colonnade Rd.
Napean, Ontario K2E 7L6
Canada
phone: 613 226 8874
fax: 613 226 8639

Fisher Scientific Co.
International Headquarters
50 Fadem Road
Springfield, NJ 07081-3193
USA
phone: 201 467 6400
fax: 201 379 7415

Fisher Scientific Singapore Pte. Ltd.
152 Beech Road
#12-05/07, The Gateway East
Singapore 0718
Singapore
phone: 65 293 8433
fax: 65 293 3525

Follansbee Steel Corp.

Sheet Metal Specialty Div.
P.O. Box 610
One State St.
Follansbee, WV 26037
USA

phone: 800 624 6906
fax: 304 527 1269

Carbon steels, sheet, strip

Follsain Wearparts Foundry

Gilmorton Road
Lutterworth, Leicestershire LE17 4HA
England

Stainless steels, cast; Cast iron

Formulated Resins Inc.

Div. of TACC International Corp.
PO Box 535
Air Station Industrial Park
Rockland, MA 02370
USA

phone: 617 878 7015
fax: 617 871 6772

Epoxies; Polyurethanes

Franklin Fibre-Lamitex Corporation

PO Box 1768
Wilmington, DE 19899
USA

phone: 302 652 3621
fax: 302 571 9754

Epoxies; Phenolics; Polyesters, thermoset

Fuchs, Otto Metallwerke

Postfach 1261
Derschlager Strasse 26
D-5882 Meinerzhagen
Germany

phone: 49 23 54 73 1
fax: 49 23 54 73 201

Aluminum and alloys; Copper and alloys;
Magnesium and alloys; Titanium and alloys

Furon

29982 Ivy Glenn Drive
Laguna Niguel, CA 92677
USA

phone: 714 831 5350
fax: 714 643 1548

Fluoropolymers, chloropolymers; Polyisoimides

Fusion Inc.

4658 E. 355th St.
P.O. Box 970
Willoughby, OH 44094
USA

phone: 216 946 3300
fax: 216 942 9083

Aluminum and alloys; Copper and alloys; Lead and alloys; Nickel, alloys, superalloys; Precious metals; Tin and alloys; Zinc and alloys

G & S Titanium, Inc.

1550 Spruce St.
Wooster, OH 44691
USA

phone: 800 860 0564
fax: 216 262 1550

Nickel, alloys, superalloys; Titanium and alloys

G.A. Avril Company

Brass and Bronze Ingot Division
4445 Kings Run Drive
P.O. Box 32066
Cincinnati, OH 45232-0066
USA

phone: 513 641 0566
fax: 513 641 0568

Copper and alloys; Lead and alloys; Tin and alloys

GAF Corp.

1361 Alps Rd.
Wayne, NJ 07470
USA

phone: 201 628 3000

Polyesters, thermoplastic

Gallagher Corp.

3966 Morrison Dr.
Gurnee, IL 60031
USA

phone: 708 249 3440
fax: 708 249 3473

Polyurethanes

GE Plastics

150 North Street
Suite 27
Pittsfield, MA 01201
USA

phone: 800 845 0600

Polycarbonates; Polyesters, thermoplastic;
Polyisoimides; Polyphenylenes; Polystyrenes

Gecamines La General De Carrieres De Mines

Lubumbashi, B. P. 450
Kinshasa B.P.8714
Republique du Zaire

Cadmium, chromium; Cobalt, alloys, superalloys; Precious metals; Zinc and alloys

GenCorp Polymer Products

350 Springside Drive
Akron, OH 44333
USA

phone: 216 668 7000
fax: 216 668 7019

Polystyrenes; Polyvinyl chlorides

OTHER LOCATIONS

GenCorp Polymer Products
Fabricated Plastics Headquarters
1722 Indian Wood Circle/Suite A
Maumee, OH 43537
USA
phone: 419 891 1500
fax: 419 891 4437

GenCorp Polymer Products
Plastic Extrusions Division
605 West Eichel Avenue
Evansville, IN 47710
USA
phone: 812 428 6400
fax: 812 428 6468

GenCorp Polymer Products
Coated Fabrics Division
133 Yorkville Road East
Columbus, MS 39702
USA
phone: 601 327 1522
fax: 601 329 7773

GenCorp Polymer Products
Plastic Extrusions Division
3333 South Zero Street
P.O. Box 6298
Fort Smith, AR 72906
USA
phone: 501 646 4535
fax: 501 648 3945

(continued)

GenCorp Polymer Products (cont'd)

GenCorp Polymer Products
Flexible Film Division
Chambers Avenue
Jeannette, PA 15644
USA
phone: 412 523 5441
fax: 412 523 7306

GenCorp Polymer Products
Latex Headquarters
165 South Cleveland Avenue
Mogadore, OH 44260
USA
phone: 216 628 6550
fax: 216 628 6501

GenCorp Polymer Products
Rigid Plastics Division
One General Street
Newcomerstown, OH 43832
USA
phone: 614 498 5900
fax: 614 498 5448

GenCorp Polymer Products
Reneer Films
Hickory Drive
P.O. Box 429
Auburn, PA 17922-0429
USA
phone: 717 366 1051
fax: 717 366 4058

General Metals Powder Co.

1195 Home Ave.
Akron, OH 44310
USA

phone: 216 633 1226
fax: 216 630 4053

Copper and alloys

General Motors Corp.

Power Train Division
MC 2J33
895 Joslyn Rd.
Pontiac, MI 48340-2920
USA

phone: 313 857 4084

Aluminum and alloys; Manganese

General Plastics Manufacturing Co.

4910 Burlington Way
PO Box 9097
Tacoma, WA 98409
USA

phone: 206 473 5000
fax: 206 473 5104

Polyurethanes

Genstar Stone Products Co.

Executive Plaza IV
11350 McCormick Rd.
Hunt Valley, MD 21031
USA

phone: 410 527 4000
fax: 410 527 4258

Fillers, additives

Geon Company

6100 Oak Tree Blvd.
Cleveland, OH 44131
USA

phone: 800 438 4366
fax: 216 447 7730

Acrylics; Polyurethanes; Polyvinyl chlorides; Thermoplastic elastomers

Georgia Iron Works Co.

Subs. of G W Industries, Inc.
5000 Wrightsboro Road
Grovetown, GA 30813
USA

phone: 706 863 1011
fax: 706 860 5897

Cast iron

Georgia Marble Co.

1201 Roberts Blvd., Bldg. 100
Kennesaw, GA 30144-3619
USA

phone: 404 421 6500
fax: 404 421 6507

Fillers, additives

Georgia-Gulf Corp.

PVC Div. Hdq.
PO Box 629
Plaquemine, LA 70765-0629
USA

phone: 504 685 1200
fax: 504 685 1205

Polyvinyl chlorides

Glacier Vandervell Ltd.

Kellmarnock, Ayrshire KA1 3NA
Scotland

phone: 44 851 936 881
fax: 44 563 721 93

Copper and alloys

Glastic Corporation

Subsidiary of Kobe Steel
4321 Glenridge Rd.
Cleveland, OH 44121-2891
USA

phone: 216 486 0100
fax: 216 486 1091

Epoxies; Polyesters, thermoset

Globe Metallurgical Sales, Inc.

6450 Rockside Woods Boulevard
Suite 390
Cleveland, OH 44131
USA

phone: 800 433 0223
fax: 216 328 9416

Ferroalloys, pure iron

Glyco-Metall-Werke

Postfach 13 03 35
Stielstrasse 11
Wiesbaden 13 D-6200
Germany

phone: 49 0 61 21 20 11

Copper and alloys; Lead and alloys; Tin and alloys

Goldman Titanium Company

Div. of GCF, Inc.
105 Dorothy St.
PO Box 246
Buffalo, NY 14240-0246
USA

phone: 716 823 9900
fax: 716 823 6886

Titanium and alloys; Zirconium

Graphite Metallizing Corp.

1050 Nepperhan Ave.
PO Box 110
Yonkers, NY 10702
USA

phone: 914 968 8400
fax: 914 968 8468

Carbon, graphite, diamond

Great Lakes Carbon Corp.

320 Old Briarcliff Rd.
Briarcliff Manor, NY 10510
USA

phone: 800 828 6601

Carbon, graphite, diamond

Great Western Steel Co.

2310 W. 58 St.
Chicago, IL 60636
USA

phone: 312 434 5800
fax: 312 434 3033

Tool steels

Grede Foundries, Inc.

9898 W. Blue Mound Rd.
P.O. Box 26499
Milwaukee, WI 53226
USA

phone: 414 257 3600
fax: 414 256 9399

Carbon steels, cast; Alloy steels, cast; Cast iron

Greene, Tweed Engineered Plastics

Div. of Greene, Tweed & Co., Inc.
PO Box 305
1880-B North Penn Road
2075 Detwiler Rd.
Kulpsville, PA 19443
USA

phone: 215 997 2555
fax: 215 997 2490

Polyketones

Greenleaf Corp.

Greenleaf Technical Ceramics Div.
P.O. Box 2420
Livermore, CA 94550
USA

phone: 510 449 3606

Engineering ceramics; High alumina and "cermets"; Carbides

(continued)

Greenleaf Corp. (cont'd)

OTHER LOCATION

Greenleaf Corp.
Composites Div.
1 Greenleaf Dr.
Saegertown, PA 16433
USA

phone: 814 763 2915
fax: 814 763 4423

Greenville Metals, Inc.

99 Crestview Drive Extension
Transfer, PA 16154
USA

phone: 412 646 0654
fax: 412 646 0661

Carbon steels, cast; Stainless steels, cast; Tool steels; Cobalt, alloys, superalloys; Magnesium and alloys; Nickel, alloys, superalloys; Refractory metals

Greenville Tube Corp.

PO Box 299
Downers Grove, IL 60515
USA

phone: 708 985 5600
fax: 708 985 6138

Stainless steels, mill shapes

GREFCO, Inc.

3435 W. Lomita Blvd.
Torrance, CA 90505
USA

phone: 310 517 0700
fax: 310 517 0794

Fillers, additives

Gulf States Steel, Inc.

174 S. 26th Street
Gadsden, AL 35904-1935
USA

phone: 800 423 0004
fax: 205 543 6768

Carbon steels, sheet, strip; Carbon steels, plate;
Alloy steels, sheet, strip; Alloy steels, plate

Gunite Corp.

302 Peoples Ave.
Rockford, IL 61104-7092
USA

phone: 815 964 3301
fax: 815 965 9197

Cast iron

Gustav Grimm Edelstahl-Werk GmbH

Postfach 15 01 36
Ronsdorfer Strasse 170-174
5630 Remscheid 1 (Haddenbach)
Germany

phone: 49 02191 4350
fax: 49 02191 43540

Alloy steels; Stainless steels; Tool steels

H.C. Starck

45 Industrial Place
Newton, MA 02164
USA

phone: 617 630 5800
fax: 617 630 5879

Refractory metals

Hackett Brass Foundry

1200 Lillibridge
Detroit, MI 48214
USA

phone: 313 822 1214
fax: 313 822 0407

Copper and alloys

Hamilton Foundry

200 Industrial Drive
Harrison, OH 45030
USA

phone: 513 367 6900
fax: 513 367 0102

Cast iron

Hamilton Technology, Inc.

101 Queen St.
P.O. Box 4787
Lancaster, PA 17604
USA

phone: 717 299 2581
fax: 717 397 8510

Alloy steels; Ferrous electronic materials; Titanium and alloys

Hammond & Irving Inc.

254 North St.
P.O. Box 656
Auburn, NY 13021
USA

phone: 315 253 6265
fax: 315 253 3136

Tool steels

Handy & Harman

Fairfield Plant
1770 Kings Highway
Fairfield, CT 06430
USA

phone: 203 259 8321
fax: 203 259 8264

Cadmium, chromium; Copper and alloys; Precious metals; Zinc and alloys

OTHER LOCATIONS

Handy & Harman
Clad Metals & Arts
P.O. Box 16055
231 Ferris Avenue
East Providence, RI 02916
USA
phone: 800 245 2728
fax: 401 434 9340

Lucas-Milhaupt, Inc.
5656 South Pennsylvania Ave.
Cudahy, WI 53110
USA
phone: 800 558 3856
fax: 414 769 1093

Handy & Harman
Refining
300 Rye Street
South Windsor, CT 06074
USA
phone: 203 282 0871
fax: 203 528 4748

Harris, Arthur & Co.

210 N. Aberdeen Street
Chicago, IL 60607
USA

phone: 312 666 6832
fax: 312 666 6893

Copper and alloys

Harris, J. W., Co., Inc.

10930 Deerfield Rd.
Cincinnati, OH 45242
USA

phone: 513 473 2006

Stainless steels; Ferrous welding, hard facing electrodes; Aluminum and alloys; Copper and alloys; Manganese; Precious metals; Zinc and alloys

OTHER LOCATIONS

Harris-Euro Corp.
Camino de la Piscina
17700 La Jonquera, Girona
Spain
phone: 34 72 555252
fax: 34 72 555380

J.W. Harris Co., Inc.
4410-E Bankers Circle
Atlanta, GA 30360
USA
phone: 404 447 9253
fax: 404 242 6649

J.W. Harris Co., Inc.
40 West Third Street
South Boston, MA 02127
USA
phone: 617 268 4450
fax: 617 268 2742

Harrison Alloys Inc.

308 Middlesex Street
P.O. Box 31
Harrison, NJ 07029
USA

phone: 201 483 4800
fax: 201 483 7100

Iron-base superalloys; Ferrous electronic materials; Copper and alloys; Nickel, alloys, superalloys

Harrison Steel Castings Co.

900 Mound St.
P.O. Box 60
Attica, IN 47918
USA

phone: 317 762 2481

Alloy steels, cast; Cast iron

Harsco Corp.

P.O. Box 8888
Camp Hill, PA 17001-8888
USA

phone: 717 763 7064
fax: 717 763 6424

Carbon steels, mill shapes; Carbon steels, plate; Alloy steels, mill shapes; Alloy steels, plate

OTHER LOCATIONS

Harsco GmbH
Postfach 1565
D-2250 Husum
Germany
phone: 49 4841 7790

Harsco U.K., Ltd.
Oxford Street
Bilston, West Midlands WV14 7EG
U.K.
phone: 44 902 409281

Harvey Titanium Ltd.

1805 Colorado Ave.
Santa Monica, CA 90404
USA

phone: 800 321 0909
fax: 310 829 7248

Titanium and alloys

Haynes International, Inc.

1020 West Park Ave.
P.O. Box 9013
Kokomo, IN 46904-9013
USA

phone: 317 456 6000
fax: 317 456 6079

Stainless steels; Iron-base superalloys; Cobalt, alloys, superalloys; Copper and alloys; Nickel, alloys, superalloys; Refractory metals

OTHER LOCATIONS

Haynes International, Inc.
Stadium Plaza
1520 South Sinclair Street
Anaheim, CA 92806
USA
phone: 714 978 1775
fax: 714 978 1743

Haynes International, Inc.
Highway 80 West
Route 1 Box 8
Arcadia, LA 71001-9701
USA
phone: 318 263 9571
fax: 318 263 8088

Haynes International, Inc.
430 Hayden Station Road
Windsor, CT 06095
USA
phone: 203 688 7771
fax: 203 688 5550

Haynes International, Inc.
The Northwood Industrial Park
12231 FM 529
Houston, TX 77041
USA
phone: 713 937 7597
fax: 713 937 4596

Haynes International, Ltd.
P.O. Box 10
Parkhouse Street
Openshaw
Manchester M11 2ER
England
phone: 44 61 230 7777
fax: 44 61 223 2412

Haynes International, S.A.R.L.
B.P. 7110
F-95054 Cergy Pontoise Cedex
France
phone: 33 1 34 48 3100
fax: 33 1 30 37 8022

Haynes International, S.R.L.
Viale Brianza 8
20127 Milano
Italy
phone: 39 2 261 41331
fax: 39 2 282 8273

Nickel Contor, AG
Gotthardstrasse 21
CH-8022 Zurich
Switzerland
phone: 41 1 201 7322
fax: 41 1 201 7333

Haysite Reinforced Plastics

An Alco Industries Company
5599 New Perry Hwy.
Erie, PA 16509
USA

phone: 814 868 3691
fax: 814 864 7803

Polyesters, thermoset

Heckford, Arthur E., Ltd.

Birmingham Metal Works
Frederick St. & Regent St.
Birmingham B1 3HH
England

phone: 44 021 236 0525

Copper and alloys; Nickel, alloys, superalloys

Hercules Inc.

Hercules Plaza
Wilmington, DE 19894
USA

phone: 800 247 4372
fax: 302 594 6909

Epoxies; Polyesters, thermoset; Fibers

(continued)

Hercules Inc. (cont'd)

OTHER LOCATION
Hercules Inc.
Bacchus Works
P.O. Box 98
Magna, UT 84044
USA

phone: 801 250 5911

Hertel Cutting Technologies

1000 Clearview Court
Oak Ridge, TN 37830
USA

phone: 800 631 8682
fax: 615 481 6058

Refractory metals

OTHER LOCATION
Hertel Cutting Technologies
141 Market Street
Kenilworth, NJ 07033
USA

Hetzel & Co. Metallhuttenwerk GmbH

Postfach 2242
Rotterdamer Strasse 135
D-8500 Nurnberg 60
Germany

phone: 49 0911 64 19 0
fax: 49 0911 64 19 15

Lead and alloys

Hexcel

Structural Products Div.
5794 W. Las Positas Blvd.
PO Box 8181
Pleasanton, CA 94588
USA

phone: 510 847 9500
fax: 510 734 9042

Aluminum and alloys; Epoxies; Phenolics; Polyimides

Himont USA Inc.

Three Little Falls Centre
2801 Centerville Rd.
PO Box 15439
Wilmington, DE 19850-5439
USA

phone: 302 996 6000
fax: 302 996 6051

Polyolefins

Hobart Bros. Co.

600 W. Main St.
Troy, OH 45373
USA

phone: 513 332 5123
fax: 513 332 5024

Ferrous welding, hard facing electrodes; Aluminum and alloys

Hobart/TAFA Inc.

146 Pembroke Rd.
Concord, NH 03301
USA

phone: 603 224 9585
fax: 603 225 4342

Nickel, alloys, superalloys; Tin and alloys; Zinc and alloys

Hoechst Celanese International

Advanced Materials Group and
High Performance Polymers
26 Main St
Chatham, NJ 07928
USA

phone: 201 635 2600
fax: 201 635 4330

Acetals; Fluoropolymers, chloropolymers; Polyamides; Polyesters, thermoplastic; Polyolefins; Polyphenylenes; Polyurethanes; Polystyrenes; Thermoplastic elastomers; Fibers

OTHER LOCATIONS

Hoechst do Brazil
Quimica e Farmaceutica S.A.
Fabrica Suzano
Suzano S.A.
Brazil
phone: 55 476 1222

Celanese Mexicana S.A.
Engineering Resins
Av. Revolucion No. 1425
Mexico 01040 D.F.
Mexico
phone: 52 525 325 5726
fax: 52 525 325 5630

Custom Compounding Inc.
50 Milton Drive
Ashton, PA 19014
USA
phone: 215 497 8899
fax: 215 494 7050

Hoechst AG
P.O. Box 80 03 20
D-6230 Frankfurt am Main 80
Germany
phone: 49 69 6600
fax: 49 69 303665

(continued)

Hoechst Celanese International (cont'd)

Hoechst Canada Inc.
Plastic Products Division
4045 Cote Vertu Blvd.
Montreal, Quebec H4R 1R6
Canada
phone: 514 871 5600
fax: 514 871 5570

Hoechst Canada Inc.
Plastic Products Department
100 Tempo Ave.
Willowdale, Ontario M2H 2NB
Canada
phone: 416 497 4700
fax: 416 495 8381

Hoechst Celanese
Advanced Materials Group
801 Brickell Ave.
Suite 923
Miami, FL 33131
USA
phone: 305 374 8226
fax: 305 372 0189

Hoechst Celanese Corp.
Rt. 202-206 North
Somerville, NJ 08876
USA
phone: 800 554 6782

Hoechst Celanese Corp.
Custom Compounding
50 Milton Drive
P.O. Box 2158
Aston, PA 19014
USA

Hoechst Celanese Corp.
13800 South Lakes Drive
Charlote, NC 28224

Hoechst Celanese Corp.
Polymer Composites
4610 Theurer Boulevard
Winona, MN 55987
USA

Hoechst Celanese Corp.
51 JFK Parkway
Short Hills, NJ 07078
USA

Hoechst Celanese International
Engineering Thermoplatics Unit
1601 West LBJ Freeway
Dallas, TX 75234
USA
phone: 214 277 3900
fax: 214 277 3910

Hoechst Celanese International
901 Mittel Drive
Wood Dale, IL 60191
USA
phone: 708 616 2100
fax: 708 616 2162

(continued)

Hoechst Celanese International (cont'd)

Hoechst Celanese International
1411 West 190th Street
Suite 235
Los Angeles, CA 90248
USA
phone: 310 516 1643
fax: 310 516 6143

Hoeshst Celanese International
1195 Centre Road
Auburn Hills, MI 48326
USA
phone: 800 962 7448
fax: 313 377 2981

Hoechst do Brazil
Quimica e Farmaceutica S.A.
Avenida das Nacoes Unidas 18001
Sao Paulo SP 047951
Brazil
phone: 55 525 7795
fax: 55 525 7738

Polymer Composites Inc.
4610 Theurer Boulevard
P.O. box 30013
Winona, MN 55987
USA
phone: 507 454 4150
fax: 507 454 6116

Hoechst Engineering and Performance Plastics Asia Pacific Ltd.
1208 Tai Yau Building
181 Johnston Road
Wanchai
Hong Kong
phone: 852 5 891 1
fax: 852 5 838 9

Hoeganaes Corp.

River Rd. & Taylors Ln.
Riverton, NJ 08077
USA

phone: 609 829 2220
fax: 609 786 2574

Carbon steels, powder; Alloy steels, powder;
Stainless steels, powder; Ferroalloys, pure iron

Hoskins Mfg. Co.

10776 Hall Rd.
P.O. Box 218
Hamburg, MI 48139-0218
USA

phone: 313 231 1900
fax: 313 231 4311

Ferrous electronic materials; Copper and alloys; Nickel, alloys, superalloys; Refractory metals

Howmet Corp.

475 Steamboat Rd.
Greenwich, CT 06836-1960
USA

phone: 203 661 4600
fax: 203 661 0422

Cobalt, alloys, superalloys; Nickel, alloys, superalloys; Titanium and alloys; Carbides

OTHER LOCATIONS

Ceramet Casting
2175 Avenue C
Bethlehem, PA 18017
USA
phone: 215 266 0270
fax: 215 266 1990

Cercast Casting
3905 Industrial Boulevard
Montreal North, Quebec H1H 2Z2
Canada
phone: 514 322 2371
fax: 514 322 1340

Ceramic Products
1600 South Warner Street
Whitehall, MI 49461
USA
phone: 616 894 5686
fax: 616 894 7512

Cercon Casting
201 Cercon Drive
Hillsboro, TX 76645
USA
phone: 817 582 3413
fax: 817 582 2466

(continued)

Howmet Corp. (cont'd)

Cercor Casting
93 Mountainview Road North
Georgetown, Ontario L7G 4J6
Canada
phone: 416 877 6936
fax: 416 877 6938

Ciral
4, avenue Laurent Cely
92606 Asieres Cedex
France
phone: 33 1 40 80 25 02/3
fax: 33 1 47 33 72 28

Ciral IT
Via P.G.A. Filippino, 119
00144 Rome
Italy
phone: 39 06 52 91 840
fax: 39 06 52 94 268

Exeter Alloy
Heron Road
Exeter, Devon EX2 7LL
England
phone: 44 392 432 751
fax: 44 392 211 774

Exeter Casting
Kestrel Way
Exeter, Devon EX2 7LG
England
phone: 44 392 70731
fax: 44 392 413147

Hampton Casting
One Howmet Drive
Hampton, VA 23661
USA
phone: 804 838 4680
fax: 804 837 1965

Howmet Corp.
Dover Alloy Division
Roy Street
Dover, NJ 07801
USA
phone: 201 361 2310
fax: 201 989 5405

Komatsu-Howmet, Ltd.
1-1, 3-Chome Ueno
Hirakata-shi
Osaka-fu 573
Japan
phone: 81 720 49 31235
fax: 81 720 49 4128

LaPorte Casting
1110 East Lincoln Way
LaPorte, IN 46350
USA
phone: 219 326 7400
fax: 219 324 3193

Sigma Casting
925 S. Charlie Road
City of Industry, CA 91748
USA
phone: 818 965 2457
fax: 818 964 5440

(continued)

Howmet Corp. (cont'd)

Ti-Cast
1600 South Warner Street
Whitehall, MI 49461
USA
phone: 616 894 5686
fax: 616 894 7512

Titanium Ingot Plant
555 Benston Road
Whitehall, MI 49461
USA
phone: 616 894 5686
fax: 616 894 7354

Whitehall Casting
One Misco Drive
Whitehall, MI 49461
USA
phone: 616 894 5686
fax: 616 894 7347

Wichita Falls Casting
6200 Central Freeway
Wichita Falls, TX 76307
USA
phone: 817 855 8100
fax: 817 855 1479

Hoyt Alloys

Div. of Hoyt Darchem Ltd.
Garth Road
Lower Morden
Surrey SM4 4LT
England

phone: 44 081 337 7744
fax: 44 081 330 4813

Bismuth and alloys; Lead and alloys; Tin and alloys

Hsimex Southland Polymers

14030 Gannet St.
Santa Fe Springs, CA 90670
USA

phone: 310 921 0444
fax: 310 921 0034

Polystyrenes

Huls America Inc.

Turner Place
Piscataway, NJ 08855-0365
USA

phone: 908 981 5000
fax: 908 981 5414

Polyamides

Postfach 1320
D-4370 Marl
Germany

phone: 49 02365 491

Huntsman Chemical Corp.

5100 Bainbridge Blvd.
Chesapeake, VA 23320
USA

phone: 804 494 2500
fax: 804 494 2602

Polyolefins; Polystyrenes

OTHER LOCATION

Huntsman Chemical Corp.
2000 Eagle Gate Tower
Salt Lake City, UT 84111
USA

phone: 801 532 5200

Hussey Copper Ltd.

100 Washington St.
Leetsdale, PA 15056
USA

phone: 412 857 4200

Copper and alloys

Hyde, A. L., Co.

1 Main Street
Grenloch, NJ 08032
USA

phone: 800 234 4933
fax: 609 232 1754

Polyamides; Polyphenylenes

Hydro Magnesium

MgMt, Postboks 25 50
Bugg 92
N 3901 Porsgrunn
Norway

phone: 47 3 56 3897

Magnesium and alloys

ICD Group Inc.

600 Madison Avenue
17th Floor
New York, NY 10022
USA

phone: 212 644 1500

Fibers

ICI Fiberite Corp.

501 West Third St.
Winona, MN 55987
USA

phone: 507 454 3611
fax: 507 454 5105

Epoxies; Phenolics; Polyesters, thermoset; Polyketones; Polyimides; Silicones

OTHER LOCATIONS

ICI Composites Inc.
2055 East Technology Circle
Tempe, AZ 85284
USA
phone: 602 730 2010

ICI Fiberite
4300 Jackson Street
Greenville, TX 75401
USA
phone: 903 457 8500

ICI Fiberite
2 Lakeside Avenue
Delano, PA 18220
USA
phone: 717 467 3315

ICI Fiberite
645 North Cypress
Orange, CA 92667
USA
phone: 714 639 2050

IMI plc

P.O. Box 216
Witton
Birmingham B6 7BA
England

phone: 44 021 356 4848
fax: 44 021 356 3526

Aluminum and alloys; Copper and alloys; Refractory metals; Titanium and alloys

(continued)

IMI plc (cont'd)

OTHER LOCATIONS

IMI Rolled Metals Ltd.
a Subsidiary of IMI plc
P.O. Box 703
Witton
Birmingham B6 7UP
England
phone: 44 021356 3344

IMI Yorkshire Alloys Ltd.
P.O. Box 166
Leeds LS1 1RD
England
phone: 44 0532 701107
fax: 44 0532 719803

IMI Titanium Ltd.

a Subsidiary of IMI plc
P.O. Box 704
Witton
Birmingham B6 7UR
England
phone: 44 021 356 1155
fax: 44 021 356 5413

Titanium and alloys

OTHER LOCATIONS

IMI Titanium Inc.
Titech Pomona Operations
PO Box 3060
4000 West Valley Blvd.
Pomona, CA 91769-3060
USA
phone: 714 595 7455
fax: 714 595 7450

IMI Titanium Inc.
Albany Activity
150 West Queen Avenue
P.O. Box 908
Albany, OR 97321-0336
USA
phone: 503 926 7711
fax: 503 967 7786

IMI-TECH Corp.

701 Fargo Ave.
Elk Grove Village, IL 60007
USA

phone: 708 981 7676
fax: 708 981 7806

Polyimides

Inco Alloys International, Inc.

P.O. Box 1958
3200 Riverside Dr.
Huntington, WV 25720
USA

phone: 304 526 5100
fax: 304 526 5441

Stainless steels; Iron-base superalloys; Ferrous welding, hard facing electrodes; Ferrous electronic materials; Aluminum and alloys; Copper and alloys; Nickel, alloys, superalloys

OTHER LOCATIONS

Wiggin Works
Holmer Road
Hereford HR4 9SL
England
phone: 44 0432 382200
fax: 44 0432 264030

Inco Alloys International, Inc.
222 N. Mountain Ave.
Suite 104
Upland, CA 91786
USA
phone: 714 946 0803
fax: 714 946 1660

Inco Alloys International Pte. Ltd.
50 Robinson Rd. 06-00
MNB Building
Singapore 0106
Singapore
phone: 65 222 3988
fax: 65 221 4298

Inco Alloys International, Inc.
Foulkwood Plaza
Suite 202
2036 Foulk Road
Wilmington, DE 19810
USA
phone: 302 529 1818
fax: 302 529 1248

(continued)

Inco Alloys International, Inc. (cont'd)

Inco Alloys International, Inc.
Suite 255
1297 Johnson Ferry Road
Marietta, GA 30068
USA
phone: 404 971 4570
fax: 404 971 4780

Inco Alloys International, Inc.
2260 Hicks Road
Suite 405
Rolling Meadows, IL 60008-1205
USA
phone: 708 255 0888
fax: 708 255 3187

Inco Alloys International, Inc.
Franklin Office Park West
38 Pond Street, Suite 108
Franklin, MA 02038
USA
phone: 508 520 3801
fax: 508 520 3805

Inco Alloys International, Inc.
23200 Chagrin Blvd.
P.O. Box 22218
Beachwood, OH 44122
USA
phone: 216 464 8705
fax: 216 464 4674

Inco Alloys International, Inc.
13231 Champion Forest Drive
Suite 200
Houston, TX 77069
USA
phone: 713 440 6600
fax: 713 440 4627

Inco Alloys International, Inc.
Tour Neptune
Cedex 20
F-92086 Paris-La-Defense
France
phone: 33 1 47 76 42 51
fax: 33 1 49 00 19 31

Inco Alloys International, Inc.
Talstrasse 32a
D-4000 Dusseldorf
Germany
phone: 49 0211 37 90 78
fax: 49 0211 37 98 64

Inco Alloys Pte Ltd.
14 Queens Road
Melbourne, Victoria 3004
Australia
phone: 61 3 867 3666
fax: 61 3 867 6872

Inco Alloys Pte. Ltd.
15 F1/Wilson House
19-27 Wyndham Street
Central
Hong Kong
Hong Kong
phone: 852 05 212 333
fax: 852 05 810 1965

Inco Pacific Ltd.
Taiwan Branch Office
5-3, 5th Floor, Union Plaza
498 Min Chuan East Rd.
Taipei
Taiwan R.O.C.

Indium Corporation of America

1676 Lincoln Ave.
Utica, NY 13503
USA

phone: 315 853 4900
fax: 315 853 1000

Bismuth and alloys; Indium; Lead and alloys; Precious metals; Tin and alloys

OTHER LOCATION

Indium Corporation of Europe

7 Grisdale Court
Woborn Road Industrial Estate
Kempston, Bedford Mk42 7EE
England

phone: 44 0234 840255
fax: 44 0234 841498

Industrial Dielectrics Inc.

PO Box 357
Noblesville, IN 46060
USA

phone: 317 773 1766
fax: 317 773 3877

Polyesters, thermoset

Industrial Steels Inc.

3561 Industrial Rd.
P.O. Box 346
Mims, FL 32754
USA

phone: 407 267 2341
fax: 407 383 9072

Stainless steels

Inland Steel Co.

30 West Monroe Street
Chicago, IL 60603
USA

phone: 312 346 0300

Carbon steels, mill shapes; Carbon steels, sheet, strip; Carbon steels, plate; Alloy steels, mill shapes; Alloy steels, sheet, strip; Alloy steels, plate

Insulation Materials Inc.

PO Box 1404
New Haven, CT 06505
USA

phone: 203 562 2171
fax: 203 562 6864

Epoxies

International Nickel Inc.

Park 80 West-Plaza Two
Saddle Brook, NY 07662
USA

phone: 201 843 8600

Alloy steels, mill shapes; Alloy steels, plate; Stainless steels, cast; Iron-base superalloys; Cast iron; Nickel, alloys, superalloys

International Polymer Corp.

5602 Armour Dr.
Houston, TX 77020
USA

phone: 713 673 7700
fax: 713 673 4488

Polyketones; Polyphenylenes

Isabellenhutte

Isabellenhutte Hewster GmbH KG
Postfach 14 53
35664 Dillenburg
Germany

phone: 49 02771 9340
fax: 49 02771 23030

Copper and alloys; Nickel, alloys, superalloys

Iten Industries

4001 Benefit Ave.
PO Box 9
Ashtabula, OH 44004
USA

phone: 216 997 6134
fax: 216 992 3614

Epoxies; Phenolics; Polyesters, thermoset

(continued)

OTHER LOCATIONS

Australian Isola Materials
Unit 5-6/86 Heathcote Rd.
P.O. Box 16
Moorebank, NSW
Australia
phone: 61 02 602 7299
fax: 61 02 602 1302

Iten Industries
10855 Hasty Avenue
Downey, CA 90241
USA
phone: 310 869 5949

Iten Industries
Regional Sales, Fabrication and Distribution Plant
P.O. Box 314
Addison, IL 60101
USA
phone: 708 543 2820
fax: 708 543 1463

Iten Industries
Rt. 1 Box 195
Sallisaw, OK 74955
USA
phone: 918 775 9323

Iten Industries
1125 E. Street
Mansfield, MA 02048
USA
phone: 508 339 2110

La Fontaine & Associates
Enrique Rebsamen 411-B
Mexico D.F. C.P. 03020
Mexico
phone: 52 525 639 6464
fax: 52 525 639 0192

Iten Industries
P.O. Box 2038
Covington, GA 30209
USA
phone: 404 786 6185

La Fontaine & Associates
R.R. #2
Ravenna, Ontario NOH 2EO
Canada
phone: 705 444 2409
fax: 705 444 6508

Japan Steel Works Ltd.

1-2 Yurakucho
1-Chome
Chiyoda-Ku
Tokyo
Japan

Alloy steels, sheet, strip; Alloy steels, plate

Jelenko, J. F., & Co.

99 Business Park Dr.
Armonk, NY 10504
USA

phone: 914 273 8600

Precious metals; Refractory metals

Jelliff Corp.

P.O. Box 758
354 Pequot Ave.
Southport, CT 06490
USA

phone: 203 259 1615
fax: 203 255 7908

Copper and alloys; Nickel, alloys, superalloys

Jessop Steel Co.

Jessop Place
500 Green Street
Washington, PA 15301
USA

phone: 412 222 4000

Alloy steels, mill shapes; Stainless steels, mill shapes; Tool steels

Jet Moulding Compounds Ltd.

405 Fairall St.
Ajax, ON L1S 1R8
Canada

phone: 416 686 1707
fax: 416 427 9403

Polyesters, thermoset

Johnson Matthey plc

Materials Technology Division
New Garden House
78 Hatton Garden
London EC1N 8JP
England

phone: 44 01 430 0011
fax: 44 01 242 6024

Copper and alloys; Lead and alloys; Precious metals; Refractory metals

OTHER LOCATION
Johnson Matthey
Materials Technology Division
1401 King Rd.
West Chester, PA 19380
USA

phone: 800 523 7792

Johnson Mfg. Co.

Div. of Gardiner Metal Co.
114 Lost Grove Road
PO Box 96
Princeton, IA 52768
USA

phone: 319 289 5123
fax: 319 289 3825

Stainless steels; Lead and alloys; Zinc and alloys

(continued)

Johnson Mfg. Co. (cont'd)

OTHER LOCATIONS

Gardiner Metal Co.
4820 South Campbell Ave.
Chicago, IL 60632
USA
phone: 312 847 0100
fax: 312 847 2314

Gardiner-West
2021 N. Glassell
Orange, CA 92665
USA
phone: 714 921 9050
fax: 714 921 2119

Johnstown Corp.

545 Central Avenue
Johnstown, PA 15902
USA

phone: 800 458 1505

Alloy steels, mill shapes; Stainless steels, mill shapes; Tool steels

Junker, Otto, GmbH

5101 Lammersdorf uber Aachen
Germany

phone: 49 24 73 601 0
fax: 49 24 73 601 221

Stainless steels; Nickel, alloys, superalloys

Kaiser Aluminum & Chemical Corp.

P.O. Box 877
6177 Sunol Blvd.
Pleasanton, CA 94566
USA

phone: 510 462 1122
fax: 510 847 4791

Aluminum and alloys

OTHER LOCATIONS

Kaiser Aluminum & Chemical Corp.
P.O. Box 619
Erie, PA 16503
USA
phone: 814 454 4571
fax: 814 456 9046

Kaiser Aluminum & Chemical Corp.
L.A. Extrusion
6250 E. Bandini Blvd.
Los Angeles, CA 90040
USA
phone: 213 726 8011

Kaiser Aluminum & Chemical Corp.
1111 Airline Hwy. 61
Gramercy, LA 70052
USA
phone: 504 869 5711

Kaiser Aluminum & Chemical Corp.
E. 2111 Hawthorne Road
Mead, WA 99021
USA
phone: 509 466 3300
fax: 509 467 1698

Kaiser Aluminum & Chemical Corp.
Greenwood Forge
1508 Hwy 246 South
Greenwood, SC 29646
phone: 803 223 6515
fax: 803 223 4575

Kaiser Aluminum & Chemical Corp.
So. Hebron Road
1459 Hebron Road
Heath, OH 43056
USA
phone: 614 522 1151
fax: 614 522 0422

Kaiser Aluminum & Chemical Corp.
Greenwood Machine
813 W. Alexander Ave.
Greenwood, SC 29646
USA
phone: 803 227 6113
fax: 803 227 6387

Kaiser Aluminum & Chemical Corp.
1001 McWane Blvd.
Oxnard, CA 93033
USA
phone: 805 488 4401
fax: 805 986 3334

(continued)

Kaiser Aluminum & Chemical Corp. (cont'd)

Kaiser Aluminum & Chemical Corp.
P.O. Box 1215
Sherman, TX 75090
USA
phone: 903 893 5566
fax: 903 894 0480

Kaiser Aluminum & Chemical Corp.
7311 E. 41st Street
Tulsa, OK 74145
USA
phone: 918 627 0100
fax: 918 663 3795

Kaiser Aluminum & Chemical Corp.
3400 Taylor Way
Tacoma, WA 98421
USA
phone: 206 383 1461
fax: 206 591 0493

Kaiser Aluminum & Chemical Corp.
4125 Mahoning Rd. N.E.
Canton, OH 44705
USA
phone: 216 456 4321
fax: 216 456 0530

Kaiser Aluminum & Chemical Corp.
E. 15000 Euclid
Spokane, WA 99215
USA
phone: 509 924 1500
fax: 509 927 6597

Kaneka America Corp.

Kanegafuchi Chemical Industry Co. Ltd.
2-4, 3-chome
Nakanoshima, Kita-ku
Osaka
Japan
phone: 81 06 208 6417
fax: 81 06 208 6401

Polyolefins

Kanthal A.B.

119 Wooster St.
Bethel, CT 06801-0281
USA

phone: 203 744 1440
fax: 203 748 2229

Ferrous electronic materials; Copper and alloys; Nickel, alloys, superalloys; Electrical ceramics

OTHER LOCATIONS

Kanthal
Stepanska 34
CS-112 17 Prague 1
Czechoslovakia
phone: 42 2 235 9565
fax: 42 2 235 2405

Kanthal Bulgaria
Mizia St 25
1124 Sofia
Bulgaria
phone: 359 2 46 54 01
fax: 359 2 44 32 48

Kanthal
ul Marszalkowska 115
Pokoj no 324
00-102 Warsaw
Poland
phone: 48 22 249 473
fax: 48 22 249 059

Kanthal Electroheat (HK) Ltd.
1407 Star House 3
Salisbury Road
Tsim-Sha-Tsui
Kowloon
Hong Kong
phone: 852 735 09 33
fax: 852 735 72 38

Kanthal A.B.
Box 501 S-734 01
Hallstahammar
SWEDEN
phone: 46 0220 21000
fax: 46 0220 13910

Kanthal Electroheat (SEA) Pte. Ltd.
73 Ayer Rajah Crescent
#01-01
Singapore 0513
Singapore
phone: 65 778 55 33
fax: 65 777 32 03

Kanthal Australia Pty Ltd.
P.O. Box 9
Smithfield, NSW 2164
Australia
phone: 61 2 604 5711
fax: 61 2 609 7514

Kanthal GmbH
Postfach 147
D-6082 Morfelden-Waldorf
Germany
phone: 49 6105 400 10
fax: 49 6105 747 90

(continued)

Kanthal A.B. (cont'd)

Kanthal Korea Ltd.
Young Dong P.O. Box 275
Kangnam-ku
Seoul
South Korea
phone: 82 2 515 1431
fax: 82 2 515 1435

Kanthal Ltd.
Inveralmond
Perth, Scotland PH1 3EE
U.K.
phone: 44 738 209 31
fax: 44 738 209 36

Kanthal Ltd.
Woodlands
Festival Way
Stoke-on-Trent, Staffs ST1 5SG
U.K.
phone: 44 782 21 96 00
fax: 44 782 20 26 06

Kanthal Norden AB
Postboks 1034
DK-2650 Hvidovre
Denmark
phone: 45 31 78 66 00
fax: 45 31 78 07 03

Kanthal Norden AB
Postboks 54
N-1472 Fjellhamar
Norway
phone: 47 67 97 29 70
fax: 47 67 97 46 65

Kanthal Norden AB
Timmermalmintie 19A
SF-01680 Vantaa
Finland
phone: 358 0 85 21 633
fax: 358 0 85 21 651

Kanthal SA
BP no 38
Zl de Colombes
F-92704 Colombes Cedex
France
phone: 33 1 478 071 61
fax: 33 1 478 197 50

Kanthal Spain SA
Carrer de l'Horta
Parcela 30, nave 7
Poligono Industrial El Pla
E-08750 Molins De Rei (Barcelona)
Spain
phone: 34 3 680 28 24
fax: 34 3 680 29 66

Kanthal Trading Australasia Ltd.
P.O. Box 27148
Mt. Roskill
Auckland
New Zealand
phone: 64 9 626 4151
fax: 64 9 626 2795

Kanthal-Gadelius KK
P.O. Box 1892
Tokyo Central 100-91
Japan
phone: 81 3 344 318 21
fax: 81 3 344 053 03

(continued)

Kanthal A.B. (cont'd)

Societa Italiana Kanthal SpA
Via LB Alberti N 49
I-20092 Cinisello-Ba
Italy

phone: 39 2 660 491 77
fax: 39 2 612 33 94

Kaydon Ring & Seal Inc.

P.O. Box 626
Baltimore, MD 21203
USA

phone: 301 547 7715

Stainless steels; Cast iron; Copper and alloys; Refractory metals

Kelly Foundry & Machine Co.

P.O. Drawer 1789
Elkins, WV 26241
USA

phone: 304 636 0390

Cast iron; Copper and alloys

Kennametal Inc.

P.O. Box 231
Latrobe, PA 15650
USA

phone: 800 533 1965

Cobalt, alloys, superalloys; Copper and alloys; Nickel, alloys, superalloys; Refractory metals; Carbides

Kind & Co Edelstahlwerk

Postfach 21 80
D-51662 Wiehl
Germany

phone: 49 02262 840
fax: 49 02262 84175

Alloy steels; Stainless steels; Tool steels

Klockner-Werke AG

Postfach 100853
D-4100 Duisburg 1
Germany

phone: 49 0203 396 1
fax: 49 0203 396 3535

Stainless steels

KM-kabelmetal AG

Postfach 33 20
Klosterstrasse 29
D-49023 Osnabruck
Germany

phone: 49 0541 321 1544
fax: 49 0541 321 1571

Aluminum and alloys; Copper and alloys

OTHER LOCATIONS

KM (Suisse) SA
Thurgauer Str. 76, P.O. Box
CH-8050 Zurich
Switzerland
phone: 41 01 302 1777
fax: 41 01 302 8196

KM-Asia Pte. Ltd.
1 Selegie Road
09-13 Paradiz Centre
Singapore 0718
Singapore
phone: 65 3 37 86 71
fax: 65 3 37 85 71

KM-Belgium b.v.b.a.
St. Bavostraat 78-80
B-2610 Antwerpen-Wil
Belgium
phone: 32 3 8 30 34 64
fax: 32 3 8 30 63 73

KM-Iberica S.L.
Zurbano 34-3 dcha
E-28010 Madrid
Spain
phone: 34 01 319 6348
fax: 34 01 3194550

KM-kabelmetal AG
P.O. Box 3320
D-49074 Osnabruck
Germany
phone: 49 05 41 321 15
fax: 49 05 41 321 15

KM-kabelmetal America Inc.
1000 Jorie Boulevard
Suite 111
Oak Brook, IL 60521
USA
phone: 708 990 2025
fax: 708 990 0258

KM-kabelmetal Danmark ApS
Landbrugsvej 8
DK-5260 Odense S
Denmark
phone: 45 65 91 64 10
fax: 45 65 91 6411

KM-kabelmetal France S.A.R.L.
Sales Office Argenteuil
141/145 Rue Michel Carre
P.O. Box 159
F-95105 Argenteuil C
France
phone: 33 01 34 26 45
fax: 33 01 30 76 29

(continued)

KM-kabelmetal AG (cont'd)

KM-kabelmetal Italia s.r.l.
Piazza Giulio Cesare 9
I-20145 Milan
Italy
phone: 39 02 48 19 37
fax: 39 02 4 81 560

KM-kabelmetal Vertriebsgesellschaft fur
Slamastr. 48
P.O. Box 128
A-1232 Vienna
Austria
phone: 43 0222 6 16 36
fax: 43 0222 6 16 36

KM-kabelmetal Portugal Lda.
Rua Brito Camacho, Lote 5C
Loja
Quinta da Figueirinha
P-2780 Oeiras
Portugal
phone: 351 01 4 42 602
fax: 351 01 441 5620

KM-kabelmetal Nederland BV
Stationsweg 4, 1e etage
P.O. Box 146
NL-3300 AC Dordrecht
Netherlands
phone: 31 078 31 1444
fax: 31 078 13 82 90

KM-kabelmetal UK Ltd.
9/17 Tuxford Road
Hamilton Industrial Park
GB-Leicester LE4 7TZ
England
phone: 44 05 33 46 11
fax: 44 05 33 46 11

Kobe Steel USA Inc.

a Subsidiary of Kobe Steel, Lt
535 Madison Ave.
New York, NY 10022
USA

phone: 212 751 9400

Carbon steels, mill shapes; Carbon steels, sheet, strip; Carbon steels, plate; Alloy steels, mill shapes; Alloy steels, sheet, strip; Alloy steels, plate; Tool steels

Krupp Stahl AG

Stainless Steel Mfr.
Alleestrasse 165
D 4630 Bochum 1
Germany

phone: 49 0234 627 1
fax: 49 0234 627 408

Alloy steels, mill shapes; Alloy steels, sheet, strip; Stainless steels, mill shapes; Stainless steels, sheet, strip; Stainless steels, plate; Tool steels

OTHER LOCATION

Krupp Stahl AG
Postfach 101220
D-5900 Siegen
Germany
phone: 49 0271 801 0
fax: 49 0271 86978

Krupp VDM

11210 Steeple Crest Dr.
#120
Houston, TX 77065-4939
USA

phone: 713 955 6683
fax: 713 890 4316

Alloy steels; Stainless steels; Stainless steels, mill shapes; Stainless steels, sheet, strip; Stainless steels, forged; Iron-base superalloys; Copper and alloys; Manganese; Nickel, alloys, superalloys; Ferrous electronic materials; Refractory metals; Titanium and alloys

OTHER LOCATIONS

Krupp VDM
P.O. Box 14155-1979
Ostad Motahari 368
15968 Teheran
Iran
phone: 98 02 89 19 17
fax: 98 02 89 96 98

Krupp VDM (Schweiz)AG
Lange Gasse 90
Postfach
CH-4002 Basel
Switzerland
phone: 41 061 2 72 84
fax: 41 061 2 72 84

(continued)

Krupp VDM (cont'd)

Krupp VDM GmbH
Plettenberger Strasse 2
P.O. Box 1820
D-5980 Werdohl
Germany
phone: 49 2392 55 0
fax: 49 23 92 55 22

Krupp VDM GmbH
Postfach 1820
D-58778 Werdohl
Germany
phone: 49 23 92 55 22
fax: 49 23 92 55 25

Krupp VDM GmbH
Tubes & Components
P.O.Box 14 17 20
D-47207 Duisburg
Germany
phone: 49 20 65 9933
fax: 49 20 65 993 42

Krupp VDM GmbH
Heilbronner Strasse 67
D-70191 Stuttgart
Germany
phone: 49 711 2 59 81
fax: 49 711 2 59 82

Krupp VDM GmbH
Wittestrasse 49
D-13509 Berlin
Germany
phone: 49 30 432 4036
fax: 49 30 435 2968

Krupp VDM GmbH
Postfach 4247
D-90022 Nurnberg
Germany
phone: 49 911 6 60 91/
fax: 49 911 66 16 48

Krupp VDM GmbH
Stubelallee 49
D-01309 Dresden
Germany
fax: 49 351 459 4910

Krupp VDM GmbH
Feldbergstrasse 15
D-79194 Gundelfingen
Germany
phone: 49 761 5 88 58
fax: 49 761 58 45 80

Krupp VDM GmbH
Friedensallee 128
D-22763 Hamburg
Germany
phone: 49 40 88 20 21
fax: 49 40 8 80 88 6

Krupp VDM GmbH
Postfach 50 01 63
D-60391 Frankfurt
Germany
phone: 49 69 58 02 154
fax: 49 69 58 02 109

(continued)

Krupp VDM (cont'd)

Krupp VDM Hongkong Ltd.
19th Floor
80 gloucester Road
Wanchai
Hong Kong
phone: 852 02 5 272 00
fax: 852 02 5 272 04

VDM Canada Ltd.
11 Allstate Parkway
Suite 203
Markham, Ontario L3R 9TS
Canada
phone: 416 477 2064
fax: 416 477 2817

S.A. VDM Belgium N.V.
Rue de la Station 150
Bte 1
B-1410 Waterloo
Belgium
phone: 32 02 354 2900
fax: 32 02 354 3626

VDM de Mexico, S.A. de C.V.
Avenida Lic, Juan Fernandez Al
No.21 1er. Piso Fraccionamient
San Pablo Xalpa, Tlalnepantla,
E C.P 54090
Mexico
phone: 52 05 3 91 06 05
fax: 52 05 3 91 37 99

VDM (U.K.) Ltd.
VDM House
111 Hare Lane
Claygate-Esher
Surrey, KT 10 OQY
England
phone: 44 0372 46 71 3
fax: 44 0372 46 63 8

VDM do Brasil Ltda.
Av. Brig. Faria Lima 1,885
15, Andar-Sala 1505/6
01451 Sao Paulo-SP
Brazil
phone: 55 011 8 14 56
fax: 55 011 8 15 92

VDM Australia Pty. Ltd.
P.O. Box 552
58 Duerdin Street
North Clayton, Victoria 3168
Australia
phone: 61 03 561 1311
fax: 61 03 561 4465

VDM Ges.m.b.H.
Representative Office
154 Rakovsky Str.
BG-1000 Sofia
Bulgaria
phone: 359 092 62 82 2

(continued)

Krupp VDM (cont'd)

VDM Gesellschaft m.b.H.
Theresianumgasse 7
Postfach 6
A-1040 Wien
Austria
phone: 43 2 22 5 04 26
fax: 43 2 22 5 05 26

VDM Gessellschaft m.b.H.
Predstavnistvo za Jugoslaviju
Zajceva 44a
HR-41000 Zagreb
Croatia and Slo
phone: 41 22 08 56
fax: 41 23 36 19

VDM Iberica S.A.
Travessera de Gracia 18, 5, 3
E-08021 Barcelona
Spain
phone: 34 3 2 00 90 11
fax: 34 3 2 0022 54

VDM Italia S.R.L.
Via Montepulciano 11
I-20124 Milano
Italy
phone: 39 02 66 7117 7
fax: 39 02 6 69 4930

VDM Japan K.K.
2nd Floor, Ochanomizu Itch Bld
3-3 Kanda-Surugadai, Chiyoda-K
Tokyo 101
Japan
phone: 81 3 32 95 45 9
fax: 81 3 32 95 45 9

VDM Korea Co., Ltd.
Room 502 Dong Nam Bldg.
997-11 Daechi 3-Dong
Kangnam-Ku
Seoul
Korea
phone: 82 02 552 6321
fax: 82 02 552 6320

VDM Nederland B.V.
Stationsweg 4
NL-3311 JW Dordrecht
Netherlands
phone: 31 078 31 69 66
fax: 31 078 31 58 57

VDM S.a.r.l.
160 Bureaux de la Colline
F-92213 Saint-Cloud
France
phone: 33 01 46 02 87
fax: 33 01 49 11 18

VDM Singapore Pte.Ltd.
8 Shenton Way
08-02 A Treasury Bldg.
Singapore 1542
Singapore
phone: 65 227 9031
fax: 65 224 8528

VDM Technologies Corp.
10 Sylvan Way
P.O. Box 362
Parsippany, NJ 07054
USA
phone: 201 267 8545
fax: 201 292 4919

(continued)

Krupp VDM (cont'd)

VDM Technology (SA)(Pty.)Ltd.
P.O. Box 84
Wendywood 2144
South Africa

phone: 27 011 444 3620

Krupp-Hoesch-Stahl AG

Postfach 10 50 42
D-44120 Dortmund
Germany

phone: 49 0231 844 597
fax: 49 0231 844 473

Alloy steels, sheet, strip

OTHER LOCATION
Krupp-Hoesch-Stahl AG
Werk Benrath
Hildener Strasse 80
D-40597 Dusseldorf
Germany

fax: 49 2 11 71 07 5

Kulite Tungsten Corp.

160 E. Union Ave.
East Rutherford, NJ 07073
USA

phone: 201 438 9000
fax: 201 438 0891

Refractory metals

Ladish Co., Inc.

5481 South Packard Ave.
Cudahy, WI 53110
USA

phone: 414 747 2611

Alloy steels, mill shapes; Tool steels

OTHER LOCATIONS

Ladish Co., Inc.
Arkansas Division
P.O. Box 1600
Russellville, AR 72801
USA

Ladish Co., Inc.
Kentucky Division
P.O. Box 159
Cynthiana, KY 41031
USA

Lanxide Corporation

Tralee Industrial Park
Newark, DE 19714-6077
USA

phone: 302 454 0244

Aluminum and alloys; Carbides; Oxides

LaSalle Steel Co.

1412 150th Street
Hammond, IN 46327-1799
USA

phone: 800 852 6003

Carbon steels, mill shapes; Alloy steels, mill shapes

Latrobe Steel Co.

2626 Ligonier St.
Latrobe, PA 15650
USA

phone: 412 537 7711
fax: 412 537 8633

Alloy steels, mill shapes; Stainless steels, mill shapes; Tool steels

Lavin, R., & Sons, Inc.

3426 S. Kedzie Ave.
Chicago, IL 60623
USA

phone: 312 847 1800
fax: 312 847 6909

Aluminum and alloys; Copper and alloys

Lenzing AG

A-4860 Lenzing
Austria

phone: 43 07672 701
fax: 43 07672 75740

Polyimides

OTHER LOCATION

Lenzing Fibers Corp.

6100 Fairview Rd.
Suite 300
Charlotte, NC 28210
USA

phone: 704 554 7699
fax: 704 554 0577

Leonische Drahtwerke AG

Leonische of America, Inc.
Marienstr. 7
W-8500 Nurnberg 1
Germany

phone: 49 0911 2023 22
fax: 49 0911 2023 23

Copper and alloys

OTHER LOCATION

Leonische of America, Inc.
301 Griffith Road
Chicopee, MA 01022
USA

phone: 413 593 6618
fax: 413 593 6639

Lesjofors Steel AB

Box 1067
S68096 Lesjefors
Sweden

phone: 46 0590 31085
fax: 46 0590 31105

Alloy steels, mill shapes; Tool steels

Lincoln Electric

Airco Filler Metal Products
22801 St. Clair
Cleveland, OH 44117-1199
USA

phone: 216 481 8100
fax: 216 486 1751

Ferrous welding, hard facing electrodes; Aluminum and alloys; Copper and alloys; Nickel, alloys, superalloys

OTHER LOCATIONS

Big Three Lincoln Alaska Inc.
6415 Arctic Boulevard
Anchorage, AK 99518-1533
USA
phone: 907 563 3133
fax: 907 564 9743

Lincoln Big Three, Inc.
1350 Choctaw
P.O. Box 3274
Baton Rouge, LA 70821-3274
USA
phone: 504 357 4331
fax: 504 357 4367

Lincoln Electric (U.K.)
Mansfield Road
Aston
Sheffield S31 OBS
England
phone: 44 742 872401
fax: 44 742 872582

Lincoln Electric C.A.
Calle 8, Edificio Lion
Piso 1, Offic. 107
Caracas, La Urbina 1070
Venezuela
phone: 011 582 241 331
fax: 011 582 241 874

Lincoln Electric Co.
22801 St. Clair Ave.
Cleveland, OH 44117
USA
phone: 216 383 2267
fax: 216 486 4649

Lincoln Electric Company of Canada Ltd.
179 Wicksteed Avenue
Toronto, Ontario M4G 2B9
Canada
phone: 416 421 2600
fax: 416 421 3065

Lincoln Electric do Brazil
Avenida Brasil 50.701
Zona Industrial De Palmares
Campo Grande
23.065 Rio De Janeir
Brazil
phone: 55 21 395 2121
fax: 55 21 395 2700

Lincoln Electric France SA
Avenue Franklin Roosevelt BP21
76120 Grand-Quevilly
France
phone: 33 35 694422
fax: 33 35 692540

(continued)

Lincoln Electric (cont'd)

Lincoln Electric Italia SRL
Via Gelasio Adamoli 239 b/c
16141 Genova GE
Italy
phone: 39 10 835 5507
fax: 39 10 835 5050

Lincoln Electric Mexicana S.A. de C.V.
Calz. Azcapotzalco La Villa No
Colonia Industrial Vallejo
Delegacion Azcapotzalco
02300, Mexico, D.F.
Mexico
phone: 011 525 567 090
fax: 011 525 567 523

Lincoln KD S.A.
Ctra. Laurea Miro, 396-398
08980 Sant Feliu de Liobregat
Barcelona
Spain
phone: 34 3 6665051
fax: 34 3 6668001

Lincoln Norweld AS
P.O. Box 25
Risoya
3291 Stavern
Norway
phone: 47 33 198 500
fax: 47 33 197 375

Lincoln Smitweld Belgium NV
Avenue Paul Gilson Laan 470
B-1620 Drogenbos
Belgium
phone: 32 2 3770071
fax: 32 2 3781877

Lincoln Smitweld BV
P.O. Box 253
6500 AG Nijmegen
Netherlands
phone: 31 80 522911
fax: 31 80 522670

Lincoln Svetshuset AB
P.O. Box 758
531 17 Lidkoping
Sweden
phone: 46 510 22045
fax: 46 510 62791

Messer Lincoln GmbH
P.O. Box 1120
D-65831 Liederbach/T
Germany
phone: 49 69 33007 0
fax: 49 69 33007 105

Nippon Lincoln Electric K.K.
Fukushima Factory
1-45, Aza-Nakamaru, Ohaza-Yama
Naraha-Machi, Futaba-Gun
Fukushima-Ken 979-05
Japan
phone: 81 240 25 5556
fax: 81 240 25 4169

The Lincoln Electric Co. (Asia Pacific)
Blk 219 Henderson Industrial P
#02-01 Henderson Road
Singapore 0315
Singapore
phone: 65 276 0878
fax: 65 276 0836

(continued)

Lincoln Electric (cont'd)

The Lincoln Electric Co. (Australia) Pty
35 Bryant Street
Sidney, Padstow NSW 2211
Australia
phone: 61 2 772 7222
fax: 61 2 792 1387

The Lincoln Electric Company
Latin America Headquarters
Miami Lakes Commerce Center
5715 N.W. 158th Street
Hialeah, FL 33014-6719
USA
phone: 305 556 0142
fax: 305 822 9129

Linderme Tube Co.

1500 East 219th St.
Cleveland, OH 44117
USA

phone: 216 481 5300
fax: 216 481 5383

Aluminum and alloys; Copper and alloys

Little Falls Alloys, Inc.

189 Caldwell Ave.
Paterson, NJ 07501
USA

phone: 201 278 1666
fax: 201 278 7345

Copper and alloys

LNP Engineering Plastics

An ICI Company
412 King St.
Malvern, PA 19355
USA

phone: 215 644 5200

Acetals; Fluoropolymers, chloropolymers; Polyamides; Polycarbonates; Polyesters, thermoplastic; Polyketones; Polyolefins; Polyphenylenes; Polystyrenes; Polystyrenes; Polysulfones

Lohmann, Friedr., GmbH

Werk fur Spezial & Edelstahle
Postfach 3262
Ruhrtal 2
D-5810 Witten-Herbed
Germany

phone: 49 023 02 70 14
fax: 49 0 23 02 70 1

Tool steels

Lord Corp.

2000 West Grandview Blvd.
PO Box 10038
Erie, PA 16514-0038
USA

phone: 814 868 3611
fax: 814 864 3452

Polyurethanes

LTV Steel Co.

25 W. Prospect Ave.
Cleveland, OH 44115
USA

phone: 216 622 5000

Carbon steels, mill shapes; Carbon steels, sheet, strip; Alloy steels, mill shapes; Alloy steels, sheet, strip

Ludlow Steel Corp.

5437 Perkins Road
Bedford, OH 44146
USA

phone: 216 232 1700

Tool steels

Lukens Steel Co.

Lukens Inc.
ARC Building
Modena Road
Coatesville, PA 19320
USA

phone: 215 383 2791

Carbon steels, plate; Alloy steels, plate; Tool steels

Lunkenheimer Co.

1500 Waverly Avenue
Cincinnati, OH 45214
USA

phone: 513 921 3400

Stainless steels, mill shapes; Copper and alloys; Nickel, alloys, superalloys

Lustro Plastics Co.

PO Box 3011
Evanston, IL 60204
USA

phone: 312 864 4050
fax: 312 864 4689

Polyesters, thermoplastic; Polyesters, thermoset

M. C. Gill Corp.

4056 Easy St.
El Monte, CA 91731
USA

phone: 818 443 6094
fax: 818 350 5880

Epoxies; Phenolics; Polyesters, thermoset

Macaulay Foundry Inc.

811 Carleton St.
Berkeley, CA 94710
USA

phone: 510 845 2911
fax: 510 845 2918

Cast iron

Magnacast Corp.

1117 East Algonquin Road
Arlington Heights, IL 60005
USA

phone: 708 437 6000
fax: 708 437 6008

Copper and alloys

Magnequench

6435 S.Scatterfield Rd.
Anderson, IN 46013
USA

phone: 317 646 5050
fax: 317 646 5060

Ferrous electronic materials

Magnesium Elektron Ltd.

Reade Mfg. Div.
100 Ridgeway Blvd.
Lakehurst, NJ 08733
USA

phone: 908 657 6451
fax: 908 657 6628

Magnesium and alloys

OTHER LOCATION

Magnesium Elektron Ltd.
P.O. Box 6
Manchester, Swinton M27 2LS
England

phone: 44 061 794 2511
fax: 44 061 728 2813

Magnetic Metals Co.

Hayes Ave at 21st St.
P.O. Box 351
Camden, NJ 08101
USA

phone: 609 964 7842
fax: 609 963 8569

Ferrous electronic materials; Nickel, alloys, superalloys

Magnetic Shield Corp.

Perfection Mica Co.
740 N. Thomas Dr.
Bensenville, IL 60106
USA

phone: 312 766 7800

Ferrous electronic materials; Nickel, alloys, superalloys

Magnolia Metal Corporation

P.O. Box 19110
One Magnolia Center
Omaha, NE 68119
USA

phone: 402 455 8760
fax: 402 455 8762

Copper and alloys; Lead and alloys; Tin and alloys

Mahle GmbH

Postfach 50 07 69, 50 07 80
Pragstasse 26-46
D-7000 Stuttgart 50
Germany

phone: 49 0711 5011

Aluminum and alloys

Manoir Electro Alloys

Taylor Street & Abbe Road
P. O. Box 4012
Elyria, OH 44035
USA

phone: 216 323 3202

Stainless steels, cast

Manville Corp.

PO Box 5108
Denver, CO 80217-5108
USA

phone: 800 654 3103
fax: 303 978 2318

Refractory ceramics

Materials Development Corp.

81 Hicks Avenue
Medford, MA 02155
USA

phone: 617 391 0400
fax: 617 391 7964

Nickel, alloys, superalloys

McGean-Rohco, Inc.

2910 Harvard Ave.
Cleveland, OH 44105-3010
USA

phone: 216 441 4900

Nickel, alloys, superalloys

McGhan NuSil Corp.

1150 Mark Ave.
Carpinteria, CA 93013
USA

phone: 805 684 8780
fax: 805 684 2365

Silicones; Thermoplastic elastomers

McKay Welding Products

101 Trade Square East
Troy, OH 45373
USA

phone: 513 332 5706
fax: 800 577 2365

Ferrous welding, hard facing electrodes

McLouth Steel Corp.

1650 W. Jefferson Avenue
Trenton, MI 48183
USA

phone: 313 285 1200
fax: 313 246 4059

Carbon steels, sheet, strip; Alloy steels, sheet, strip;
Alloy steels, plate; Stainless steels, sheet, strip

MCP Systems, Inc.

511 Commerce Dr.
Fairfield, CT 06430-5541
USA

phone: 203 367 7761
fax: 203 368 4787

Bismuth and alloys; Cadmium, chromium; Cobalt, alloys, superalloys; Indium; Lead and alloys; Manganese; Nickel, alloys, superalloys; Refractory metals; Tin and alloys; Zinc and alloys; Thermoplastic elastomers

OTHER LOCATIONS

HEK Medizintechnik GmbH
Kanichenborn 24-28
P.O. Box 1832
D-2400 Luebeck
Germany
phone: 49 0451 5300444
fax: 49 0451 530450

MCP Metelec
Paterson Road
Finedon Road Industrial Estate
Wellingborough, Northants NN8 4BZ
U.K.
phone: 44 0933 225766
fax: 44 0933 227814

MCP France SA
5 Boulevard Jules Ferry
75011 Paris
France
phone: 33 1 47 008809
fax: 33 1 47 003795

Metalspecialties Inc.
515 Commerce Dr.
Fairfield, CT 06430
USA
phone: 203 384 0331
fax: 203 368 4082

Meridian Steel Co., Inc.

50 Jericho Turnpike
Jericho, L.I., NY 11753
USA

phone: 516 334 8250
fax: 516 334 8653

Tool steels

Metal Carbides Inc.

6001 Southern Blvd.
P.O. Box 3749
Youngstown, OH 44513
USA

phone: 216 788 6541
fax: 216 788 8506

Carbides

Metallwerk Goslar GmbH & Co. KG

Postfach 2129
Im Schleeke 8
D-38611 Goslar
Germany

phone: 49 05321 754 23
fax: 49 05321 754222

Lead and alloys

Metco Perkin-Elmer

1101 Prospect Ave.
Westbury, Long Islan, NY 11590-0201
USA

phone: 516 334 1300
fax: 516 338 2414

Stainless steels; Aluminum and alloys; Copper and alloys; Nickel, alloys, superalloys; Refractory metals; Tin and alloys; Zinc and alloys; Carbides

MG Industries

Welding Products
N94 W14355 Garwin Mace Dr.
Menomonee Falls, WI 53051-1628
USA

phone: 414 255 5520
fax: 414 255 5542

Ferrous welding, hard facing electrodes; Aluminum and alloys; Copper and alloys; Stainless steels, powder; Nickel, alloys, superalloys; Precious metals; Refractory metals

MicroMet Technology Inc.

P.O. Box 471906
Charlotte, NC 28247
USA

phone: 704 846 0522
fax: 704 846 0723

Ferroalloys, pure iron

Micron Metals Inc.

7186 West Gates Ave.
Salt Lake City, UT 84120
USA

phone: 801 250 5919
fax: 801 250 7295

Aluminum and alloys; Titanium and alloys

Middelburg Steel and Alloys (Pty) Ltd.

P.O. Box 781815
Sandton 2146, Transvaal
South Africa

phone: 011 27 783 2060/83
fax: 011 27 883 1416

Alloy steels, sheet, strip; Alloy steels, plate; Stainless steels, sheet, strip; Stainless steels, plate

Miles Inc.

Polysar Rubber Division
2603 W. Market St.
Akron, OH 44313-4270
USA

phone: 216 836 0451
fax: 216 836 0200

Polyesters, thermoplastic; Polystyrenes

OTHER LOCATION

Miles Inc.
Mobay Road
Pittsburgh, PA 15205-9741
USA

phone: 412 777 2000
fax: 412 777 2447

Miller Thermal Technologies, Inc.

P.O. Box 1081
Appleton, WI 54912
USA

phone: 414 731 6884
fax: 414 734 2160

Refractory metals

(continued)

Miller Thermal Technologies, Inc. (cont'd)

OTHER LOCATION
Alloys International
1901 Ellis School Rd.
Baytown, TX 77521
USA

phone: 713 426 5535
fax: 713 426 7484

Milne, A., & Co.

3157 Process Drive
Norcross, GA 30071
USA

phone: 404 876 0391
fax: 404 662 0357

Tool steels

Mitsubishi Materials Corp.

1-6-1, Ohtemachi
Chiyoda-ku, Tokyo 100
Japan

Alloy steels

OTHER LOCATION
Mitsubishi Yuka America, Inc.
375 Park Ave.
Suite 2209
New York, NY 10152
USA

phone: 212 980 9650
fax: 212 308 9132

(continued)

Molecu-Wire Corp.

P.O. Box 495
Farmingdale, NJ 07727
USA

phone: 201 922 9400

Copper and alloys; Nickel, alloys, superalloys

Monmouth Plastics, Inc.

814 Asbury Ave.
PO Box 921
Asbury Park, NJ 07712
USA

phone: 800 526 2820
fax: 908 775 9068

Polyolefins; Polystyrenes

Monsanto Plastics

ADC-Applications Development C
800 Worcester Street
Springfield, MA 01151
USA

phone: 413 730 2204
fax: 413 730 3521

Polyamides; Polystyrenes

Morton International

2700 E. 170th Street
Lansing, IL 60438
USA

fax: 708 868 7487

Acrylics; Polyurethanes

OTHER LOCATIONS
Morton International
Thermoplastic Polyurethane Div
137 Folly Mill Rd.
Seabrook, NH 03874
USA

Motor Castings Co.

1323 S. 65th St.
Milwaukee, WI 53214-0996
USA

phone: 414 476 1434
fax: 414 476 2845

Cast iron

MRC Polymers, Inc.

1716 West Webster Ave.
Chicago, IL 60614
USA

phone: 312 276 6345
fax: 312 276 4431

Polyamides; Polycarbonates; Polyesters, thermoset

Mueller Brass Co.

2199 LaPeer Avenue
Port Huron, MI 48060
USA

phone: 313 987 7770
fax: 313 987 7321

Aluminum and alloys; Copper and alloys; Zinc and alloys

Mukand Ltd.

Lal Bahadur Shastri Marg
Kurla
Bombay 400 070
India

phone: 91 022 511 5080
fax: 91 22 5113445

Carbon steels, mill shapes; Alloy steels, mill shapes; Stainless steels, mill shapes

Multicore

1751 Jay Ell Drive
Richardson, TX 75081
USA

phone: 214 238 1224

Lead and alloys; Tin and alloys

Nachi Fujikoshi Corp.

World Trade Center Bldg.
2-4-1 Hamamatsucho
Minato-ku
Tokyo 105
Japan

phone: 81 03 435 5139
fax: 81 03 436 2747

Alloy steels, mill shapes

National Castings, Inc.

1414 East Broadway St.
Toledo, OH 43605
USA

phone: 419 698 4341

Alloy steels, cast

National Forge Co.

100 Front St.
Irvine, PA 16329
USA

phone: 814 563 7522

Alloy steels, mill shapes

National Machinable Carbide, Inc.

415 Patricia Drive
Warminster, PA 18974
USA

phone: 215 672 4225
fax: 215 674 4013

Carbides

National Starch and Chemical Corp.

PO Box 6500
Finderne Ave.
Bridgewater, NJ 08807
USA

phone: 908 685 5000

Polyimides; Polyisoimides

Neste Polymers, Inc.

RR 2
274A
Port Murray, NJ 07865
USA

phone: 908 850 6200
fax: 201 850 7282

Polyolefins

Newcomer Products, Inc.

Box 272
Rt. 982 North
Latrobe, PA 15650
USA

phone: 412 694 8100
fax: 412 694 8620

Carbides

OTHER LOCATION
Wilmington Carbide, Inc.

2957 North Kerr Avenue
Wilmington, NC 28405
USA

phone: 919 762 1711

Newport Adhesives and Composites Inc.

1822 Reynolds Ave.
Irvine, CA 92714
USA

phone: 714 253 5680
fax: 714 253 5692

Epoxies; Phenolics

Newport Steel Corp.

Ninth and Lowell Streets
Newport, KY 41072
USA

phone: 606 292 6820

Carbon steels; Ferrous electronic materials

Ney, The J. M., Co.

Ney Industrial Park
Bloomfield, CT 06002
USA

phone: 203 242 2281

Precious metals

NGK Metals Corp.

P.O. Box 13367
Reading, PA 19612-3367
USA

phone: 215 921 5000
fax: 215 921 5358

Aluminum and alloys; Copper and alloys; Nickel, alloys, superalloys

OTHER LOCATIONS

NGK Metals Corp.
21412 Protecta Drive
Elkhart, IN 46516
USA
phone: 800 348 7483
fax: 219 522 2180

NGK Metals Corp.
11724 Willake Street
Santa Fe Springs, CA 90670
USA
phone: 800 322 4463
fax: 310 948 5837

NGK Metals Corp.
725 East Avenue
Athens, TN 37303
USA
phone: 800 423 7717
fax: 615 745 6874

Nippert Co.

801 Pittsburgh Dr.
Delaware, OH 43015
USA

phone: 614 363 1981

Copper and alloys

Nippon Stainless Steel Co. Ltd.

Kashima Works #2
Aza Hikari Ooaza Hikari
Kashima-Cho, Kashima, Ibaraki-Ken
Japan

phone: 011 81 299 8437
fax: 011 81 299 8256

Stainless steels, cast; Stainless steels, mill shapes; Stainless steels, sheet, strip

Nippon Steel U.S.A. Inc.

a subsidiary of Nippon Steel C
900 N. Michigan Avenue
Suite 2810
Chicago, IL 60611
USA

phone: 312 335 5237
fax: 312 751 0345

Carbon steels, sheet, strip; Alloy steels, mill shapes; Alloy steels, sheet, strip; Alloy steels, plate; Ferrous electronic materials

NKK Corp.

450 Park Ave.
New York, NY 10022
USA

phone: 212 826 6250

Alloy steels, sheet, strip; Alloy steels, plate; Stainless steels, sheet, strip; Stainless steels, plate; Ferrous electronic materials

OTHER LOCATION

NKK Corp.
1-2, 1-Chome, Marunouchi
Chiyoda-ku
Tokyo 100
Japan

phone: 81 03 212 7111
fax: 81 03 222 2811

Norplex Oak Inc.

Part of Allied Signal Technology
230 N. Front Street
PO Box 1448
La Cross, WI 54602-1448
USA
phone: 608 784 6070
fax: 608 784 7753

Epoxies; Fluoropolymers, chloropolymers; Phenolics; Polyesters, thermoset; Polyimides; Polystyrenes

Northwest Alloys, Inc.

P.O. Box 115
Addy, WA 99101
USA

phone: 509 935 3300/33

Magnesium and alloys

Norton Chemplast

150 Dey Rd.
Wayne, NJ 07470
USA

phone: 201 696 4700

Fluoropolymers, chloropolymers

Novacor Chemicals Ltd.

P.O. Box 2535
Station "M"
Calgary, Alberta T2P 3P7
Canada

phone: 403 290 8977
fax: 403 264 6012

Polyolefins; Polystyrenes; Silicones

(continued)

Novacor Chemicals Ltd. (cont'd)

OTHER LOCATIONS

Genesis Polymers
2550 Busha Highway
Marysville, MI 48040
USA
phone: 800 627 1221
fax: 313 364 4670

Novacor Chemicals Inc.
Suite 1400
1450 East American Lane
Schaumburg, IL 60173
USA
phone: 708 330 4479
fax: 708 330 4485

Novacor Chemicals (Canada) Ltd.
Highway 40 at Moore Road 6 (Sh
Moore Township, Ontario N0N 1M0
Canada
phone: 519 862 2961
fax: 519 862 1011

Novacor Chemicals Ltd.
Alberta Gas Ethylene-Joffre
P.O. Box 5006
Read Deer, Alberta T4N 6A1
Canada
phone: 403 342 8611
fax: 403 342 8608

Novacor Chemicals (Canada) Ltd.
P.O. Box 3042 (Mailing)
Sarnia, Ontario N7T 7M1
Canada

Novacor Chemicals, Inc.
690 Mechanic Street
Leominster, MA 01453
USA
phone: 800 626 2377
fax: 800 437 3786

Novacor Chemicals (Canada) Ltd.
Chimie Novacor Ltee.
Suite 460
3100 Cote Vertu
Saint-Laurent, Quebec H4R 2J8
Canada
phone: 514 745 1414
fax: 514 745 0874

Novacor Technical Centre
3620-32nd Street NE
Calgary, Alberta T1Y 6G7
Canada
phone: 403 291 8444
fax: 403 291 0493

Novacor Chemicals (Canada) Ltd.
2nd Floor
6711 Mississauga Road
Mississauga, Ontario L5N 2W3
Canada
phone: 416 542 3338
fax: 416 542 8075

Shincor Silicones, Inc.
1030 Evans Avenue
Akron, OH 44305
USA
phone: 216 630 9460
fax: 216 630 2857

Nuclear Metals, Inc. (NMI)

2229 Main Street
Concord, MA 01742
USA

phone: 508 369 5410
fax: 508 369 4045

Cast iron; Cobalt, alloys, superalloys; Nickel, alloys, superalloys; Refractory metals; Titanium and alloys; Zirconium

NVF Company

P.O. Box 516
Kennett Square, PA 19348-0516
USA

phone: 215 444 2800

Epoxies; Polyimides

NYCO

NYCO Minerals Inc.

124 Mountain View Drive
Willsboro, NY 12996-0368
USA

phone: 518 963 4262

Fillers, additives

(continued)

NYCO (cont'd)

OTHER LOCATION
NYCO Minerals Inc.-Europe
Ondrupuej 24
P.O. Box 88
DK-2920 Chanlo Henlu
Denmark

phone: 31 64 33 70
fax: 31 64 37 10

Oaks Foundry Inc.

P.O. Box 871
Warren, OH 44482
USA

phone: 216 372 4010

Copper and alloys

Occidental Chemical Corp.

PVC Resins and Compounds Div.
P.O. Box 809050
Dallas, TX 75244
USA

phone: 214 404 3800
fax: 214 404 3669

Polyvinyl chlorides

Olin Brass

Division of Olin Corporation
427 North Shamrock St.
East Alton, IL 62024-1174
USA

phone: 618 258 2000
fax: 618 258 3084

Copper and alloys

Olin Brass, Indianapolis

Division of Olin Corp.
P.O. Box 51519-T
1800 South Holt Rd.
Indianapolis, IN 46251-0519
USA

phone: 317 244 2461

Copper and alloys

Oregon Metallurgical Corp.

530 W. 34th St.
PO Box 580
Albany, OR 97321-0177
USA

phone: 800 547 4104
fax: 503 967 8669

Titanium and alloys; Zirconium; High alumina and "cermets"

Oregon Steel Mills Inc.

1000 SW Broadway
Suite 2200
Portland, OR 97205-3003
USA

phone: 503 240 5252
fax: 503 240 5291

Carbon steels, plate; Alloy steels, plate

Osram Sylvania Inc.

Chemical & Metallurgical Division
Hawes St.
Towanda, PA 18848-0504
USA

phone: 717 265 2121

Stainless steels, sheet, strip; Ferrous welding, hard facing electrodes; Ferrous electronic materials; Copper and alloys; Precious metals; Refractory metals; Carbides

Ossenberg & Cie, W.

Postfach 7044
5990 Altena 7
Germany

phone: 0 23 52 70 08 0
fax: 0 23 52 7 3138

Alloy steels; Stainless steels; Tool steels

Oster, A.J., Foils, Inc.

2081 McCrea St.
Alliance, OH 44601
USA

phone: 800 621 2561

Aluminum and alloys; Copper and alloys

OTHER LOCATIONS

A. J. Oster Company
Brass Division
445 Industrial Drive
Warwick, RI 02886
USA
phone: 401 739 0800
fax: 401 739 1278

A.J. Oster Company
180 Alexandra Way
Carol Stream, IL 60188
USA
phone: 800 323 9644
fax: 708 260 1102

A.J. Oster Caribe, Inc.
Building No. 18
Centro Industrial Corujo
Hato Tejas
Bayamon, Puerto Rico 00619
USA
phone: 809 785 8065
fax: 809 740 2995

A.J. Oster Company
Corporate Offices
The Summit East, #300
300 Centerville Road
Warwick, RI 02886
USA
phone: 401 736 2600

A.J. Oster Company
One Frost Bridge Road
Watertown, CT 06795
USA
phone: 203 274 5441
fax: 203 274 0275

A.J. Oster Foils, Inc.
2081 McCrea Street
Alliance, OH 44601-2793
USA
phone: 800 321 9750
fax: 216 823 1705

A.J. Oster Company
6690 Grant Way
Allentown, PA 18106
USA
phone: 800 345 3060
fax: 215 481 9155

A.J. Oster West, Inc.
22833 La Palma Avenue
Yorba Linda, CA 92687-4767
USA
phone: 800 933 5933
fax: 714 692 1225

Otis Elevator Co.

521 5th Ave.
New York, NY 100175
USA

phone: 212 557 5700

Stainless steels; Cast iron; Tool steels; Aluminum and alloys

Outokumpu

129 Fairfield Way
Bloomingdale, IL 60108
USA

phone: 708 307 1300
fax: 708 980 4290

Stainless steels; Stainless steels, mill shapes; Stainless steels, sheet, strip; Cadmium, chromium; Copper and alloys; Nickel, alloys, superalloys; Zinc and alloys

Outokumpu American Brass

70 Sayre Street
P.O. Box 981
Buffalo, NY 14240
USA

phone: 716 879 6743
fax: 716 879 6899

Copper and alloys

(continued)

Outokumpu American Brass (cont'd)

OTHER LOCATIONS

Outokumpu American Brass
Franklin Operations
P.O. Box 539
Franklin, KY 42135-0539
USA
phone: 800 844 8203
fax: 502 586 7103

Outokumpu Copper Kenosha
Electrostrip
1420 63rd Street
Kenosha, WI 53143-4495
USA
phone: 414 657 5111
fax: 414 657 2399

Ovako Steel

500 Wallace Way
P.O. Box 263
York, SC 29745
USA

phone: 803 684 1737
fax: 803 684 1743

Carbon steels, mill shapes; Alloy steels, mill shapes; Stainless steels, mill shapes

Owens Corning Fiberglass Corp.

Industrial Operating Division
Fiberglas Tower
Toledo, OH 43659
USA

phone: 800 342 3745

Fibers

P.Q. Corporation

PO Box 2370
Chattanooga, TN 37409
USA

phone: 615 629 7160

Ceramics; Mica

Pan-Metallgesellschaft Baumgartner GmbH

Postfach 1024 36
Mannheim 68024
Germany

phone: 06 21 423 03 0
fax: 06 21 4230333

Copper and alloys

Paragon Steel & Tool Co.

339 Bergen Ave.
Kearny, NJ 07032
USA

phone: 201 997 1676

Tool steels

Pechiney

475 Steamboat Rd.
Greenwich, CT 06836-1960
USA

phone: 203 661 4600

Carbon steels; Alloy steels; Stainless steels; Aluminum and alloys; Copper and alloys; Magnesium and alloys; Nuclear, heavy metals, thorium; Titanium and alloys; Engineering ceramics; Carbides; Fillers, additives

Pechiney Electrometallurgie

Tour Manhattan-Cedex 21
92087 Paris La Defen
France

phone: 33 1 47 62 88 0
fax: 33 1 47 74 72 8

Ferroalloys, pure iron; Nuclear, heavy metals, thorium

Peckovers, Ltd.

2195 Langstaff Road
Concord, Ontario L4K 2B2
Canada

phone: 416 798 7325

Cast iron

Pelton Casteel, Inc.

148 West Dewey Place
Milwaukee, WI 53207-0916
USA

phone: 414 481 3400

Carbon steels; Alloy steels

Peninsular Steel Co.

P.O. Box 5040
Southfield, MI 48086
USA

phone: 313 371 9400

Tool steels

Pennsylvania Steel Corp.

12380 Beech Daly
Detroit, MI 48239
USA

phone: 313 937 3970

Tool steels

Pettibone Corp.

Beardsley & Piper Division
5501 W. Grand Ave.
Chicago, IL 60639
USA

phone: 312 237 3700

Tool steels

Pfizer Minerals, Pigments & Metals Div.

235 East 42nd St.
New York, NY 10017
USA

phone: 800 421 6337

Fillers, additives

Phelps Dodge Corp.

2600 No. Central Ave.
Phoenix, AZ 85004-3014
USA

phone: 602 234 8100

Copper and alloys

Phifer Wire Products

PO Box 1700
Tuscaloosa, AL 35403
USA

phone: 800 633 5955
fax: 205 759 4450

Aluminum and alloys

Philex Mining Corporation

Philex Bldg.
(Corner Brixton & Fairlane Sts.)
PO Box 1201, MCC, Makati
Metro Manila
Philippines

phone: 63 2 673 1971

Copper and alloys; Precious metals

Philips Elmet

Philips Elmet/Philips Lighting Co.
Div. of North American Philips
1560 Lisbon Road
Lewiston, ME 04240
USA

phone: 207 784 3591
fax: 207 786 8924

Refractory metals

Pierce & Stevens Chemical Corp.

Microtech Division
PO Box 1092
Buffalo, NY 14240
USA

phone: 716 856 4910

Fluoropolymers, chloropolymers

Plaskolite, Inc.

PO Box 1497
Columbus, OH 43216
USA

phone: 800 848 9124

Acrylics

Plastic Materials Co. Inc.

638 S. Marshall St.
PO Box 959
Milford, DE 19963
USA

phone: 302 422 3021

Acetals; Polyamides; Polycarbonates; Polyolefins; Polystyrenes

Plastic Suppliers, Inc.

2400A Marilyn Lane
Columbus, OH 43219
USA

phone: 614 471 9100
fax: 614 471 9033

Polystyrenes

Plastics Engineering Co.

3518 Lakeshore Rd.
PO Box 758
Sheboygan, WI 53082-0758
USA

phone: 414 458 2121
fax: 414 458 1923

Phenolics; Polyesters, thermoset

Plastics Manufacturing Co.

PO Box 769045
Dallas, TX 75376-9045
USA

phone: 214 330 8671
fax: 214 337 7428

Ureas

Plymouth Tube Co.

Extruded Products Div.
PO Box 45
29 W 150 Warrenville Road
Warrenville, IL 60555
USA
phone: 800 323 9506
fax: 312 393 3551

Carbon steels, mill shapes; Alloy steels, mill shapes; Stainless steels, mill shapes; Nickel, alloys, superalloys; Titanium and alloys

PM Hochtemperatur-Metall GmbH

Postfach 144
A-6600 Reutte
Austria
phone: 43 0 56 72 2248
fax: 43 0 56 72 70 5

Iron-base superalloys; Ferrous electronic materials; Aluminum and alloys; Cobalt, alloys, superalloys; Copper and alloys; Nickel, alloys, superalloys; Refractory metals; Titanium and alloys; High alumina and "cermets"; Carbides

Polaroid Corp.

Chemicals Division
238-T South Main St.
Assonet, MA 02702
USA
phone: 508 675 0808

Thermoplastic elastomers

Polifil Inc.

1112 River St.
PO Box 509
Woonsocket, RI 02895
USA

phone: 401 767 2700
fax: 401 767 2823

Polyolefins

Polyclad

40 Industrial Park Drive
P.O. Box 6299
West Franklin, NH 03235-6299
USA

phone: 603 934 5462
fax: 603 934 2670

Epoxies; Polyimides

OTHER LOCATION

Polyclad

45 Tannery St.
Franklin, NH 03235
USA

phone: 603 934 5642
fax: 603 934 5827

Polycom Huntsman Inc.

90 West Chestnut St.
Washington, PA 15301
USA

phone: 800 538 3149

Polyolefins

Polygon Co.

Industrial Park
PO Box 176
Walkerton, IN 46574-0176
USA

phone: 219 586 3122
fax: 219 586 7336

Polyallomers

Polymer Composites Inc.

PO Box 30010
4610 Theurer Blvd.
Winona, MN 55987
USA

phone: 507 454 4150
fax: 507 454 6116

Polyamides; Polyesters, thermoplastic; Polyolefins; Polyphenylenes; Polyurethanes

Polymer Extruded Products

297 Ferry St.
Newark, NJ 07105
USA

phone: 201 344 2700
fax: 201 344 7952

Acrylics

Polymer Service Corp.

37020 Garfield, Suite T-2A
Mt.Clemens, MI 48043
USA

phone: 313 263 9430

Polyamides

Polymer Technologies

414 Olive St.
Santa Barbara, CA 93101
USA

phone: 805 962 8656

Silicones

Polyply, Inc.

1540 Marion St.
Grand Haven, MI 49417
USA

phone: 616 842 6330
fax: 616 842 5320

Epoxies; Polyesters, thermoset

Pose-Marre Edelstahlwerk GmbH

Postfach 1252
D-4006 Erkrath
Germany

phone: 49 0211 24030
fax: 49 0211 2403222

Stainless steels, cast

Powder Alloy Corporation

7805 Hartford Hill Lane
Cincinnati, OH 45242
USA

phone: 513 984 4016

Carbon steels, powder; Stainless steels, powder; Aluminum and alloys; Cobalt, alloys, superalloys; Copper and alloys; Manganese; Nickel, alloys, superalloys; Precious metals; Refractory metals; Titanium and alloys; Refractory ceramics; High alumina and "cermets"; Carbides

Powdertech Corporation

5103 Evans Ave.
Valparaiso, IN 46383
USA

phone: 219 462 4141

Ferrous electronic materials; Cobalt, alloys, superalloys; Copper and alloys; Electrical ceramics

PPG Industries Inc.

Fiber Glass Products
One PPG Place
33N
Pittsburgh, PA 15272
USA

phone: 412 434 3131

Polyesters, thermoplastic; Polyolefins

PQ Corporation

Specialty Chemicals Div.
PO Box 840
Valley Forge, PA 19482
USA

phone: 215 293 7200

Fillers, additives

Precision Industries, Inc.

Relco/Sharpaloy Manufacturing
17-21 Gordon Avenue
Providence, RI 02905
USA

phone: 401 231 5700
fax: 401 941 8555

Tool steels

Precision Rolled Products, Inc.

14255 Mt. Bismark St.
P.O. Box 60010
Reno, NV 89506
USA

phone: 702 972 0272
fax: 702 972 4368

Stainless steels; Ferrous electronic materials; Cobalt, alloys, superalloys; Nickel, alloys, superalloys; Titanium and alloys

Precision Tube Co.,Inc.

287 Wissahickon Ave.
North Wales, PA 19454-9966
USA

phone: 215 699 5801
fax: 215 699 0761

Aluminum and alloys; Copper and alloys

Precision-Kidd Steel Co.

Drawer D
Aliquippa, PA 15001
USA

phone: 412 378 0084

Tool steels

Premix Inc.

Box 281
Route 20 and Harmon Road
North Kingsville, OH 44068
USA

phone: 216 224 2181
fax: 216 224 2766

Epoxies; Polyesters, thermoset

OTHER LOCATIONS

Croslene Chemical Industries Ltd.
11F, No. 22 Nanking W. Rd.
Taipei
Taiwan
phone: 886 2 555 6661
fax: 886 2 558 5135

Dia Premix Company Ltd.
3-19 Kyobashi
2-Chome, Chuo-ku
Tokyo 104
Japan
phone: 81 3 3245 8772
fax: 81 3 3245 8791

Globell Chemical Company
621 Aljunied Road
#07-05 Lipo
Factory Bldg.
Singapore 1438
Singapore
phone: 65 475 3451
fax: 65 473 9064

Permali-Premix Ltd.
125 Bristol Road
Gloucester GL 15TT
England
phone: 44 452 528671
fax: 44 452 507409

(continued)

Premix Inc.

Pianfei-Ipa S.P.A.
Via Cuneo, 27
12080 Pianfei (CN)
Italy
phone: 39 174 559511
fax: 39 174 685588

Premix Inc. Cray Valley
Tour Beau
20 rue Jean Jaures
92807 Puteaux
France
phone: 33 1 42 91 68 4
fax: 33 1 42 91 69 9

Premix Inc.
Injection Molding Division
304 Triport Road
Georgetown, KY 40324
USA
phone: 502 868 0090
fax: 502 868 0222

Premix/E.M.S. Inc.
Corporate Office
P.O. Box 2002
Commerce Road
Lancaster, OH 43130
USA
phone: 614 654 6620
fax: 614 654 6720

Premix Inc.
Lancaster Plant
2109 Commerce Road
P.O. Box 2002
Lancaster, OH 43130
USA
phone: 614 654 6620
fax: 614 654 6720

Vitro Fibras S.A.
Av. Acueducto 459
Col. Zacateno
07360 Mexico D.F.
Mexico
phone: 52 5 227 6717
fax: 52 5 586 5196

Premix Inc.
1701 West McDonald St.
P.O. Box 548
Hartford City, IN 47348-0548
USA
phone: 317 348 5780
fax: 317 348 5780

President Titanium & Steel Co.,Inc.

243 Franklin St.(Rt. 27)
PO Box 36
Hanson, MA 02341
USA

phone: 800 225 0304

Titanium and alloys

PresMet Corp.

112 Harding St.
Worcester, MA 01604
USA

phone: 508 792 6400
fax: 508 792 6427

Copper and alloys; Lead and alloys

Products Research & Chemical Corp.

PO Box 1800
5454 San Fernando Rd.
Glendale, CA 91209
USA

phone: 818 240 2060

Epoxies; Polyurethanes

Protective Metal Alloys, Inc.

8726 Production Ave., Suite A
San Diego, CA 92121
USA

phone: 800 762 9327
fax: 619 271 1790

Cobalt, alloys, superalloys; Copper and alloys; Nickel, alloys, superalloys

Prym-Werke, William, GmbH & Co. KG

Non-Ferrous Metals Div.
Postfach 1740
Zweifaller Strasse 130
D-5190 Stolberg, Rheinland
Germany
phone: 49 2402 14 1
fax: 49 2402 81841

Copper and alloys; Zinc and alloys

Pyramid Steel Company

18400 Miles
P.O. Box 28333
Cleveland, OH 44128-0333
USA

phone: 216 475 9400

Alloy steels, mill shapes; Tool steels

Pyron Corporation

a Subsidiary of Zemex Corp.
P.O. Box 310
5950 Packard Rd.
Niagara Falls, NY 14304-0310
USA

phone: 716 285 3454

Ferroalloys, pure iron; Copper and alloys

QIT-Fer et Titane Inc.

770 Sherbrooke St. W., Suite 1
Montreal, Quebec H3A 1G1
Canada

phone: 514 288 8400
fax: 514 286 9336

Cast iron

Quanex Corporation

1900-T W. Loop S
Suite 1500
Houston, TX 77027
USA

phone: 713 961 4600

Alloy steels, mill shapes

Quantum Chemical Corp.

USI Division
11500 Northlake Dr.
Cincinnati, OH 45249
USA

phone: 513 530 6500

Nuclear, heavy metals, thorium

Quinn, K.J., & Co., Inc.

135 Folly Mill Road
Seabrook, NH 03874
USA

phone: 617 321 3200

Polyurethanes

Rankin Mfg. Co.

2285 Main St.
San Diego, CA 92154
USA

phone: 619 575 6000
fax: 619 575 6003

Ferrous welding, hard facing electrodes; Cobalt, alloys, superalloys; Carbides

Raymark Friction Co.

123 E. Stiegel St.
PO Box 1050
Manheim, PA 17545
USA

phone: 717 665 2211

Phenolics

RBC Industries Inc.

80 Cypress St.
P.O.Box 8340
Warwick, RI 02888
USA

phone: 401 941 7000
fax: 401 941 0150

Epoxies; Thermoplastic elastomers

Reactor Experiments, Inc.

1275 Hammerwood Ave.
Sunnyvale, CA 94089-2231
USA

phone: 408 745 6770
fax: 408 745 7013

Precious metals

Reading Alloys, Inc.

P.O. Box 53
Robesonia, PA 19551
USA

phone: 215 693 5822
fax: 215 693 5542

Aluminum and alloys; Cadmium, chromium; Refractory metals; Vanadium; Zirconium

Reichhold Chemicals Inc.

Reactive Polymers Div.
PO Box 13582
RTP, NC 27709-3582
USA

phone: 800 874 1807

Epoxies; Polyesters, thermoplastic; Polyesters, thermoset

Remystahl

Hagener Gusstahlwerke Remy GmbH
P.O. Box 1340
D-58013 Hagen 1
West Germany

phone: 49 23 31 387 0
fax: 49 23 31 387 29

Stainless steels; Tool steels; Ferrous electronic materials; Nickel, alloys, superalloys

Renewal Service Inc.

20 Meridian Road
Eatontown, NJ 07724
USA

phone: 908 389 5050
fax: 908 389 5058

Copper and alloys

Renosol Corp.

PO Box 1424
Ann Arbor, MI 48106
USA

phone: 313 429 5418

Polyurethanes

OTHER LOCATIONS

Renosol Corp.
505 Hoover Street
Farwell, MI 48622
USA
phone: 517 588 6181

Renosol Corp.
200 Industrial Blvd.
Mauldin, SC 29622
USA
phone: 813 288 2461

Resinoid Engineering Corp.

7557 N. St. Louis Ave.
Skokie, IL 60076
USA

phone: 312 673 1050

Phenolics

Revere Copper Products, Inc.

Rome Division
P.O. Box 300
Rome, NY 13440
USA

phone: 315 338 2554
fax: 315 338 2224

Copper and alloys; Zinc and alloys

OTHER LOCATIONS
Revere Copper Products, Inc.
New Bedford Division
P.O. Box B-975
New Bedford, MA 02741
USA

phone: 800 831 1801
fax: 508 990 2954

Rex Buckeye Co. Inc.

13251 Lakewood Heights Blvd.
Cleveland, OH 44144
USA

phone: 216 671 4000

Tool steels

REXENE

5005 LBJ Freeway
Occidental Tower
Dallas, TX 75244
USA

phone: 214 450 9000

Polyolefins

Reynolds Metals Co.

Aluminum Products
P.O. Box 27003
6601 West Broad St.
Richmond, VA 23261
USA

phone: 804 281 2000
fax: 804 281 4116

Aluminum and alloys

RMI Company

1000 Warren Ave.
Niles, OH 44446
USA

phone: 216 544 7633
fax: 216 544 7796

Titanium and alloys

OTHER LOCATIONS

RMI Titanium Company
2950 Birch Street
Brea, CA 92621
USA
phone: 714 524 99112
fax: 714 579 0110

RMI Titanium Company
Keys House
Granby Avenue/Garretets Green
Birmingham B33 OSP
England
phone: 44 21 786 2121
fax: 44 21 786 2995

Rogers Corp.

One Technology Dr.
Rogers, CT 06263
USA

phone: 203 774 9605

Epoxies; Phenolics; Polyesters, thermoset; Polyimides

OTHER LOCATIONS

Rogers Corp.
Molding Materials Division
Box 550
Mill and Oakland Streets
Manchester, CT 06040
USA
phone: 203 646 5500
fax: 203 646 5503

Rogers Corp.
Power Distribution Division
5750 E. McKellips Rd.
Mesa, AZ 85205
USA
phone: 602 830 3370
fax: 602 396 6401

Rogers Corp.
Circuit Materials Division
100 N. Dobson Road
Chandler, AZ 85224
USA
phone: 602 963 4584
fax: 602 786 8256

Rogers Corp.
Composite Materials Division
One Technology Drive
Rogers, CT 06263
USA
phone: 203 774 9605
fax: 203 774 1973

Rogers Corp.
Microwave Division
100 South Roosevelt Ave.
Chandler, AZ 85226
USA
phone: 602 961 1382
fax: 602 961 4533

Rogers Corp.
Poron Materials Division
245 Woodstock Road
East Woodstock, CT 06244
USA
phone: 203 774 9605
fax: 203 928 7843

Rogers Corp.
Soladyne Division
7447 Convoy Court
San Diego, CA 92111
USA
phone: 619 279 7872
fax: 619 277 9105

Rogers Corp.
Willimantic Division
730 Windham Road
South Windham, CT 06266
USA
phone: 203 423 6323
fax: 203 423 6120

Rohm and Haas Co.

Box 219
Bristol, PA 19007
USA

phone: 800 523 1532
fax: 215 592 3377

Acrylics

Rohr & Stolberg GmbH

Postfach 9007
4150 Krefeld 12 (Lin
Germany

phone: 49 21 51 58920
fax: 49 21 51 5002 7

Lead and alloys; Zinc and alloys

Rostone

PO Box 7497
2450 Sagamore Parkway South
Layfayette, IN 47903
USA

phone: 317 474 2421
fax: 317 474 8785

Polyesters, thermoset

Rouge Steel Co.

3001 Miller Road
PO Box 1639
Dearborn, MI 48121-1639
USA

phone: 800 241 1580
fax: 313 322 4093

Carbon steels, sheet, strip; Alloy steels, sheet, strip

RTP Co.

PO Box 5439
580 East Front St.
Winona, MN 55987
USA

phone: 800 433 4787
fax: 507 454 8130

Acetals; Fluoropolymers, chloropolymers; Polyamides; Polycarbonates; Polyesters, thermoplastic; Polyketones; Polyisoimides; Polyolefins; Polyphenylenes; Polyurethanes; Polystyrenes; Polysulfones; Thermoplastic elastomers

Rypac Aluminum Recycling Ltd.

11849 Tannery Road
Surrey, B.C. V3V 3T7
Canada

phone: 604 580 7471
fax: 604 580 0255

Aluminum and alloys; Copper and alloys

S.K.F. Industries Inc.

1100 First Avenue
King of Prussia, PA 19406
USA

phone: 215 265 1900
fax: 215 265 1457

Alloy steels, mill shapes; Tool steels

Saarstahl AG

Bismarckstrasse 57-59
Postfach 10 19 80
6620 Volklingen
Germany

phone: 49 0 68 98 101
fax: 49 6898 104015

Tool steels

Saginaw Bearing Co.

1400 Agricola Drive
Saginaw, MI 48604
USA

phone: 517 752 3169

Copper and alloys

Samuel Whittar, Inc.

20001 Sherwood Ave.
Detroit, MI 48234
USA

phone: 313 893 5000
fax: 313 893 8422

Carbon steels, sheet, strip

Sanderson Kayser Ltd.

Newhall Road
Sheffield S9 2SD
England

phone: 44 0742 449 994
fax: 44 0742 430171

Alloy steels, mill shapes; Stainless steels, mill shapes; Tool steels

Sandusky International

615 West Market St.
P.O. Box 5012
Sandusky, OH 44871-8012
USA

phone: 419 626 5340

Cast iron; Copper and alloys

Sandvik Hard Materials Ltd.

Wear Parts Division
P.O. Box 89
Torrington Ave.
Coventry CV4 9XG
England

phone: 44 0203 465 911
fax: 44 0203 463 933

Carbides

Sandvik Steel Co.

Welding, Wire & Tube Divisions
P.O. Box 1220
Scranton, PA 18501
USA

phone: 717 587 5191
fax: 717 587 0747

Stainless steels, mill shapes; Stainless steels, sheet, strip; Titanium and alloys; Carbides

OTHER LOCATIONS

A/S Sandviken
Abildager 26-28
DK-2605 Brondby
Denmark
phone: 45 42 96 65 22
fax: 45 42 96 60 81

Avesta A/S
Postbox 6337
Etterstad
N-0666 Oslo 6
Norway
phone: 47 02 64 7100
fax: 47 02 64 4830

Sandvik Aciers
B.P. 6227
F-45062 Orleans, Cedex 2
France
phone: 33 38 41 41 41
fax: 33 38 63 66 17

Sandvik AG
Limmatstrasse 2
CH-8957 Spreitenbach
Switzerland
phone: 41 056 70 61 11
fax: 41 056 71 54 80

(continued)

Sandvik Steel Co. (cont'd)

Sandvik Australia Pty Ltd.
P.O. Box 23
Smithfield, N.S.W. 2164
Australia
phone: 61 02 609 0711
fax: 61 02 609 6230

Sandvik in Austria Ges.m.b.H.
Postfach 90
A-1211 Wien 21
Austria
phone: 43 1 277 37
fax: 43 1 277 37 225

Sandvik Canada Inc.
6835 Century Avenue
Mississauga, Ontario L5N 2L2
Canada
phone: 416 826 8900
fax: 416 826 0724

Sandvik Inc. Headquarters
P.O. Box 432
Dalton, PA 18414
USA
phone: 717 563 3237
fax: 717 563 2249

Sandvik de Mexico S.A. de C.V.
Apartado Postal 512
Centro
06000 Mexico, D.F.
Mexico
phone: 52 05 397 2144
fax: 52 05 397 8881

Sandvik Ireland Ltd.
Unit 30
Stillorgan Industrial Park
Stillorgan, County Dublin
Ireland
phone: 44 01 95 20 52
fax: 44 01 95 37 25

Sandvik Espanola S.A.
Apartado 92
28820 Coslada (Madri
Spain
phone: 34 91 627 1100
fax: 34 91 627 1194

Sandvik Italia S.P.A.
Casella Postale 16155
I-20160 Milano
Italy
phone: 39 02 307051
fax: 39 2 33402641

Sandvik GmbH
Postfach 10 44 51
D-4000 Dusseldorf
Germany
phone: 49 0211 50270
fax: 49 0211 502154

Sandvik K.K.
2-2-9 Murotani 2-Chome
Nishi-ku
Kobe 65122
Japan
phone: 81 078 992 0992
fax: 81 078 992 0990

(continued)

Sandvik Steel Co. (cont'd)

Sandvik Ltd.
Manor Way
Halesowen, West Midlands B62 8QZ
England
phone: 44 021 550 4700
fax: 44 021 501 3934

Sandvik Milford
P.O. Box 817
Branford, CT 06405
USA
phone: 203 481 4281
fax: 203 483 1384

Sandvik Nederland B.V.
P.O. Box 248
NL-3100 AE Schiedam
Netherlands
phone: 31 010 208 0208
fax: 31 010 437 7207

Sandvik Portuguesa Lda
Apartado 15
Venda Nova
P-2700 Amadora Codex
Portugal
phone: 351 01 490 2020
fax: 351 01 490 1574

Sandvik Process Systems
29 Commerce Way
Totowa, NJ 07512
USA
phone: 201 812 9750
fax: 201 812 8437

Sandvik Rhenium Alloys
P.O. Box 245
Elyria, OH 44036
USA
phone: 216 365 7388
fax: 216 366 9831

Sandvik Rock Tools Inc.
P.O. Box 40402-0402
Houston, TX 77240
USA
phone: 713 460 6200
fax: 713 460 6229

Sandvik S.A. (Belgium) N.v.
Chaussee de Louvain 1188
B-1200 Bruxelles
Belgium
phone: 32 02 7204930
fax: 32 02 7260369

Sandvik Saws & Tools
P.O. Box 2036
Scranton, PA 18501
USA
phone: 717 341 9500
fax: 717 341 9294

Sandvik Seamco
P.O. Box 17167
Louisville, KY 40217-1120
USA
phone: 502 636 1414
fax: 502 636 1491

(continued)

Sandvik Steel Co. (cont'd)

Sandvik South East Asia Pty Ltd.
P.O. Box 32
Jurong Town 9161
Singapore
phone: 65 265 22 77
fax: 65 264 11 78

Sandvik Special Metals
P.O. Box 6027
Kennewick, WA 99336
USA
phone: 509 586 4131
fax: 509 582 3552

Sandvik Stal Forsaljnings AB
Box 12
S-164 40 Kista
Sweden
phone: 46 08 793 0500
fax: 46 08 752 7858

Sandvik Steel
Strip Products Div.
2235 Dewey Rd.
PO Box 1287
Benton Harbor, MI 49022-1287
USA
phone: 616 926 7241
fax: 616 926 2718

Sandvik Steel AB
S-81181 Sandviken
Sweden
phone: 46 26 263322
fax: 46 26 26 40 80

Sandvik Windsor
P.O. Box 470
Milan, TN 38358
USA
phone: 901 686 9023
fax: 901 686 1303

Soumen Sandvik OY
Postbox 52
SF-01511 Vanda
Finland
phone: 358 90 87 0661
fax: 358 90 87 06 62

Sanyo Special Steel Co., Ltd.

Technical Division
3007, Nakashima, Shikama-ku
Himeji-shi, Hyogo-ke 103
Japan

phone: 81 0792 35 6111
fax: 81 0792 34 8571

Alloy steels, mill shapes; Stainless steels, mill shapes; Stainless steels, powder; Tool steels; Ferrous electronic materials; Cobalt, alloys, superalloys; Magnesium and alloys; Nickel, alloys, superalloys

OTHER LOCATIONS

Sanyo Special Steel Co., Ltd.
New York Office
445 Park Avenue
New York, NY 10022
USA
phone: 212 935 9033
fax: 212 980 8838

Sanyo Special Steel Co., Ltd.
Export Department
Nittetsu Nihonbashi Building
1-13-1 Nihonbashi, Chuo-ku
Tokyo 103
Japan
phone: 81 03 3278 8323
fax: 81 03 3278 8330

Saudi Basic Industries Corp. (SABIC)

C/O Burson-Marsteller
230 Park Avenue South
New York, NY 10003
USA

phone: 212 614 4920

Polyolefins

(continued)

Saudi Basic Industries Corp. (SABIC)

OTHER LOCATIONS

Saudi Basic Industries Corp.
3803-9 Edinburgh Tower
The Landmark, 15 Queen's Road
Central Hongkong
Hong Kong
phone: 852 5 249167/8/9

Saudi Basic Industries Corp.
Portland House
Stag Place
London SW1E 5DA
U.K.
phone: 44 8286967/8/9

Saudi Basic Industries Corp.
P.O. Box 59090
Riyadh 11525
Saudi Arabia
phone: 966 01 401 2033

Saudi Basic Industries Corp.
P.O. Box 5105
Riyadh 11422
Saudi Arabia
phone: 966 01 401 2033

Scharr Industries Inc.

40 East Newberry Rd.
Bloomfield, CT 06002-808
USA

phone: 203 243 0343
fax: 203 242 7499

Cellulosics; Polyamides; Polyesters, thermoset; Polyolefins

Schmidt & Clemens GmbH & Co.

Edelstahlwerk
Postfach Kaiserau 11 40
D-5253 Lindlar
Germany

phone: Lindlar (STD) 2

Alloy steels, cast; Alloy steels, mill shapes; Stainless steels, cast; Stainless steels, mill shapes; Tool steels

Schwaebische Huettenwerke GmbH

7080 Aslen-Wasseralf
Germany

Carbon steels, mill shapes

Schwarzkopf Technologies

Marketing Div.
35 Jeffrey Ave.
Holliston, MA 01746
USA

phone: 508 429 6801
fax: 508 429 1029

Refractory metals

Scientific Alloys Inc.

379 Old Hopkinton Rd.
P.O. Box 523
Westerly, RI 02891
USA

phone: 401 596 4947
fax: 401 596 6626

Alloy steels; Stainless steels; Cast iron; Ferrous electronic materials; Nickel, alloys, superalloys

SCM Metal Products Inc.

2601 Weck Drive
Research Triangle Pa, NC 27709
USA

phone: 919 544 8090
fax: 919 544 7996

Copper and alloys

OTHER LOCATION

N.V. SCM Europa S.A.

141 Rue Saint Lambert
B-1200 Brussels
Belgium

phone: 32 2 771 2110
fax: 32 2 771 4902

Secon Metals Corp.

7 Intervale St.
White Plains, NY 10606
USA

phone: 914 949 4757
fax: 914 949 4834

Aluminum and alloys; Lead and alloys; Precious metals

Semi-Alloy, Inc.

888 S. Columbus Ave.
Mt. Vernon, NY 10550
USA

phone: 914 664 2800
fax: 914 664 5354

Lead and alloys

Seneca Wire & Mfg. Co.

319 South Vine_Street
P.O. Box B
Fostoria, OH 44830
USA

phone: 419 435 9261
fax: 419 435 9265

Carbon steels, mill shapes

Sharon Tube Co.

PO Box 492
134 Mill St.
Sharon, PA 16146
USA

phone: 800 245 8115
fax: 412 983 1031

Carbon steels, mill shapes

Sheffield Forgemasters Ltd.

P.O. Box 108
The Old Rectory, School Hill
Whiston, Rotherham S60 1QR
England

phone: 44 0709 828233
fax: 44 0709 367765

Stainless steels, mill shapes; Stainless steels, sheet, strip; Iron-base superalloys; Ferrous electronic materials

OTHER LOCATIONS

British Rollmakers (China) Ltd.
Suite 1501, 15/F, Tower II
China HK City
33 Canton Road
Kowloon
Hong Kong
phone: 852 735 7113
fax: 852 736 3050

Euro SFM Edelstahl GmbH
Paulusstrasse 1
Postfach 140216
400 Dusseldorf 1 BRD
Germany
phone: 49 0211 682866
fax: 49 0211 682165

Forgemasters Inc.
Suite 520
363 North Sam Houston Parkway
Houston, TX 77060
USA
phone: 713 931 7385
fax: 713 820 6175

R.B. Tennent Ltd.
Whifflet Foundry
Coatbridge ML5 4RR
Scotland
phone: 44 0236 422561
fax: 44 0236 29048

(continued)

Sheffield Forgemasters Ltd. (cont'd)

Sheffield Forgemasters (Canada) Ltd.
Suite 525
155 James Street South
Hamilton, Ontario L8P 2A4
Canada

phone: 416 528 0036
fax: 416 528 2923

Sheffield Steel Corp.

PO Box 727
Industry Ave.
Joliet, IL 60434
USA

phone: 815 723 9335
fax: 815 740 4907

Tool steels

Shell Chemical Co.

One Shell Plaza
PO Box 2463
Houston, TX 77252-2463
USA

phone: 713 241 6161
fax: 713 241 8196

Polyesters, thermoset; Polyimides; Polyolefins

Shellcast Corp.

Subsidiary of Cooper Alloy Corp.
100 Bloy St.
Hillside, NJ 07205
USA

phone: 201 688 4120
fax: 201 686 9314

Stainless steels, cast; Nickel, alloys, superalloys

Sherritt Gordon Mines Ltd.

PO Box 28
Commerce Court West
Toronto, Ontario M5L 1B1
Canada

phone: 416 363 9241
fax: 416 363 1624

Cobalt, alloys, superalloys; Nickel, alloys, superalloys

Shieldalloy Metallurgical Corp.

12 West Boulevard
P.O. Box 768
Newfield, NJ 08344
USA

phone: 609 692 4200
fax: 609 692 4017

Aluminum and alloys; Magnesium and alloys; Manganese; Nickel, alloys, superalloys; Refractory metals; Titanium and alloys; Vanadium; Zirconium

(continued)

Shieldalloy Metallurgical Corp. (cont'd)

OTHER LOCATIONS

Frankel Metal Company
19300 Filer Avenue
Detroit, MI 48234
USA
phone: 313 366 5300
fax: 313 366 5305

Shieldalloy Metallurgical Corp.
818 North Mountain Avenue
Suite 215
Upland, CA 91786
USA
phone: 909 946 6600
fax: 909 949 6169

Shieldalloy Metallurgical Corp.
Suite 217
3100 Lorna Road
Birmingham, AL 35216
USA
phone: 205 823 7811
fax: 205 823 7930

Shieldalloy Metallurgical Corp.
9 Parkway Center
Suite 214
875 Greentree Road
Pittsburgh, PA 15220
USA
phone: 412 921 7900
fax: 412 921 2352

Shieldalloy Metallurgical Corp.
837 East 162nd Street
Route #6
South Holland, IL 60473
USA
phone: 708 331 3500
fax: 708 331 3517

Sidmar N.V.

John Kennedylaan 51
B-9042 Gent
Belgium

phone: 32 091 42 3111
fax: 32 091 42 3244

Carbon steels, sheet, strip; Alloy steels, sheet, strip; Ferrous electronic materials

Sierracin/Intrex

Div. of Sierracin/Sylmar Corp.
12780 San Fernando Rd.
PO Box 4700
Sylmar, CA 91342
USA
phone: 818 362 6711
fax: 818 367 5927

Polyesters, thermoset

Sifbronze

Suffolk Works
Stowmarket
Suffolk IP14 1EY
England
phone: 0449 612183

Alloy steels; Stainless steels; Aluminum and alloys; Copper and alloys; Magnesium and alloys; Precious metals

Sigmund Cohn Corp.

121 So. Columbus Ave.
Mount Vernon, NY 10553
USA
phone: 914 664 5300
fax: 914 664 5377

Precious metals

Sivyer Steel Corp.

Riverside Division
225 South 33rd Street
Bettendorf, IA 52722
USA

phone: 319 355 1811
fax: 319 355 3946

Carbon steels, cast; Alloy steels, cast; Stainless steels, cast

Slater Steel Corporation

Sorel Forge Inc.

Subsidiary of Slater Steels
P.O. Box 520
Sorel, QUE J3P 5P2
Canada

phone: 514 746 4030
fax: 514 746 4092

Stainless steels, mill shapes; Iron-base super-alloys; Tool steels; Nickel, alloys, superalloys

Solvay Polymers Inc.

PO Box 27328
Houston, TX 77227
USA

phone: 800 527 5419
fax: 713 522 2435

Fluoropolymers, chloropolymers

(continued)

Solvay Polymers Inc. (cont'd)

OTHER LOCATION

Solvay S.A.
33 rue du Prince Albert
1050 Bruxelles
Belgium

phone: 32 2 516 61 11
fax: 32 2 509 69 18

Soundcoat

1 Burt Dr.
Deer Park, NY 11729-5701
USA

phone: 516 242 2200
fax: 516 242 2246

Lead and alloys; Fibers

Southwire Company

P.O. Box 1000
Carrollton, GA 30119
USA

phone: 800 289 3833
fax: 404 832 4929

Aluminum and alloys

Spartech-Resin Exchange

One Plastics Ave.
PO Box 1719
Nash Road Industrial Park
Cape Girardeau, MO 63702-1719
USA

phone: 314 334 3434

Acrylics; Polyamides; Polystyrenes

Spaulding Composites Co.

310 Wheeler St.
Tonawanda, NY 14150
USA

phone: 716 692 2000
fax: 716 692 4410

Epoxies; Phenolics; Polyesters, thermoset

Special Metals Corp.

4317 Middle Settlement Rd.
New Hartford, NY 13413
USA

phone: 315 798 2900
fax: 315 798 2001

Iron-base superalloys; Nickel, alloys, superalloys; Refractory metals

(continued)

Special Metals Corp. (cont'd)

OTHER LOCATIONS

Special Metals Corp.
2181 El Camino Real
Suite 301
Oceanside, CA 92054
USA
phone: 619 439 9828
fax: 619 439 5720

Special Metals Corp.
Udimet Powder Division
U.S. Route 62W
Box 140
Princeton, KY 42445
USA
phone: 502 365 9551

Special Metals Corp.
2310 South Industrial Highway
Ann Arbor, MI 48104
USA
phone: 313 665 2770
fax: 313 665 7349

Udimet Special Metals LTD.
Misbourne House
Chiltern Hills
Chalfont
St. Peter, Bucks SL9 9UH
England
phone: 44 735 888 188
fax: 44 753 888 189

Spectape Inc.

PO Box 75125
Cincinnati, OH 45275
USA
phone: 606 283 2044
fax: 606 283 2048

Polyesters, thermoplastic; Polyesters, thermoset; Polyvinyl chlorides

Spectro Alloys Corp.

13220 Doyle Path
Rosemont, MN 55068
USA

Aluminum and alloys

SPS Technologies

Aerospace Products Div.
Highland Ave.
Jenkintown, PA 19046
USA

phone: 215 572 3000
fax: 215 572 3193

Carbon steels, cast; Cobalt, alloys, superalloys; Nickel, alloys, superalloys

St. Lawrence Steel Co.

2500 Crane Center Drive
Streetsboro, OH 44241
USA

phone: 216 562 9000

Tool steels

Stackpole Magnet Division

700 Elk Ave.
Kane, PA 16735
USA

phone: 814 837 7000
fax: 814 837 0277

Ferrous electronic materials; Carbon, graphite, diamond

Stahlwerk Stahlschmidt GmbH & Co.

Wiesenstrasse 30
Postfach 19 01 20

D4000 Dusseldorf 11
Germany
phone: 49 54051

Alloy steels, mill shapes; Stainless steels, mill shapes; Stainless steels, sheet, strip; Tool steels

Stahlwerke R. & H. Plate

D5880 Ludenscheid-Pl
Germany

phone: 49 02351 439 1

Alloy steels, mill shapes; Stainless steels, mill shapes; Tool steels; Refractory metals

Stainless Foundry & Engineering

5150 N. 35th Street
Milwaukee, WI 53209
USA

phone: 414 462 7220
fax: 414 462 7303

Stainless steels, cast; Nickel, alloys, superalloys; Refractory metals

Steelmark-Eagle & Globe

ANI Corporation Limited
Manchester Road (West)
Auburn, NSW 2144
Australia

phone: 61 02 749 3000
fax: 011 61 2 749 26

Carbon steels, mill shapes; Carbon steels, sheet, strip; Alloy steels, mill shapes; Alloy steels, sheet, strip; Stainless steels, mill shapes; Stainless steels, sheet, strip; Iron-base superalloys; Tool steels

OTHER LOCATIONS

Steelmark-Eagle & Globe
P.O. box 149
Nundah
Brisbane, Queensland QLD 4012
Australia
phone: 61 07 260 9222
fax: 61 07 267 8115

Steelmark-Eagle & Globe
P.O. Box 318
Springvale, Victoria 3171
Australia
phone: 61 03 549 2666
fax: 61 03 547 5776

Steelmark-Eagle & Globe
P.O. Box 152
Cardiff, NSW 2285
Australia
phone: 61 049 54 8011
fax: 61 049 56 5440

Steelmark-Eagle & Globe
GPO box 2251
Adelaide, SA 5001
Australia
phone: 61 08 268 9999
fax: 61 08 268 4199

(continued)

Steelmark-Eagle & Globe (cont'd)

Steelmark-Eagle & Globe
P.O. Box 53
Hamilton Hill, WA 6163
Australia

phone: 61 09 337 3044
fax: 61 09 337 8607

Stelco Steel

Stelco Tower
P.O. Box 2030
Hamilton, L8N 3T1 ONT
Canada

phone: 416 528 2511

Alloy steels, mill shapes; Alloy steels, sheet, strip; Alloy steels, plate

Sterling International Technology Ltd.

Colliery Lane
Exhall, Coventry CV7 9NW
England

phone: 44 203 645252
fax: 44 0203 645312

Cast iron; Aluminum and alloys; Magnesium and alloys

Stoody Company

101 S. Hanley Road
St. Louis, MO 63105
USA

phone: 314 721 5573
fax: 314 862 5803

Ferrous welding, hard facing electrodes; Nickel, alloys, superalloys

Stulz Sickles Steel Co.

P.O. Box 273
Elizabeth, NJ 07207
USA

phone: 800 351 1776
fax: 201 351 8231

Ferrous welding, hard facing electrodes; Tool steels

Sulzer Plasma Technik, Inc.

Rigackerstrasse 16
CH-5610 Wohlen
Switzerland

phone: 41 057 22 02 12
fax: 41 057 22 70 52

Aluminum and alloys; Cobalt, alloys, superalloys; Copper and alloys; Nickel, alloys, superalloys; Refractory metals; Titaniums

(continued)

Sulzer Plasma Technik, Inc. (cont'd)

OTHER LOCATION

Sulzer Plasma Technik, Inc.
1972 Meijer
Troy, MI 48084
USA

phone: 313 585 2400
fax: 313 288 4162

Sumitomo Metal America Inc.

a subsidiary of Sumitomo Metal
420 Lexington Ave.
New York, NY 10017
USA

phone: 212 949 4760

Carbon steels, mill shapes; Carbon steels, sheet, strip; Carbon steels, plate; Alloy steels, mill shapes; Alloy steels, sheet, strip; Alloy steels, plate; Stainless steels, mill shapes; Stainless steels, plate; Stainless steels, plate

Superform USA Inc.

P.O. Box 5375
Riverside, CA 92517-5375
USA

phone: 714 351 4100

Aluminum and alloys

(continued)

Superform USA Inc. (cont'd)

OTHER LOCATION

Superform USA Inc.
Regional Office
Suite 300
12444 Powerscourt Dr.
St. Louis, MO 63131
USA

phone: 314 984 6833
fax: 314 984 6832

Superior Die Casting Corp.

1001 London Road
Cleveland, OH 44110
USA

phone: 216 481 3050

Aluminum and alloys; Zinc and alloys

Superior Graphite Co.

120 S. Riverside Plaza
Chicago, IL 60606
USA

phone: 312 559 2999
fax: 312 559 9064

Carbon, graphite, diamond

OTHER LOCATION

Superior Graphite Co.
120 S. Riverside Plaza
Chicago, IL 60606
USA

phone: 312 559 2999
fax: 312 559 9064

Surahammars Bruks AB

Box 201
S-73500 Surahammar
Sweden

phone: 46 220 345 00
fax: 46 220 30 372

Tool steels; Ferrous electronic materials

OTHER LOCATIONS

CorMag Inc.

3415 Landmark Road
Burlington, Ontario L7M 1T4
Canada
phone: 416 335 3490
fax: 416 336 3477

NorMag Inc.

250 Bishop Avenue
Bridgeport, CT 06610
USA
phone: 203 335 9921
fax: 203 367 9850

SWEPCO Tube Corp.

One Clifton Blvd.
Clifton, NJ 07015
USA

phone: 201 778 3000
fax: 201 778 9289

Stainless steels, mill shapes

Tata Iron & Steel Co., Ltd.

101 Park Avenue
New York, NY 10178
USA

phone: 212 557 7979

Carbon steels, mill shapes; Alloy steels, mill shapes; Ferrous electronic materials

OTHER LOCATION

Tata Iron & Steel Co., Ltd.
43 Chowringhee Rd.
Calcutta 16
India

phone: 91 33 448301

Technetics Corp.

1600 Industrial Drive
DeLand, FL 32724-9990
USA

phone: 904 736 1700
fax: 904 738 4533

Stainless steels; Nickel, alloys, superalloys

Techni-Cast Corp.

11220 South Garfield Avenue
South Gate, CA 90280
USA

phone: 310 923 4585

Stainless steels, cast; Aluminum and alloys;
Copper and alloys; Nickel, alloys, superalloys

Technical Plastic Extruders Inc.

574 Ferry St.
Newark, NJ 07105
USA

phone: 201 589 5800

Polyvinyl chlorides

Telcon Metals Ltd.

Napier Way
Crawley
Sussex RH10 2RB
England

phone: 44 02935 28800
fax: 44 0293 536904

Ferrous electronic materials; Cobalt, alloys, superalloys; Copper and alloys; Nickel, alloys, superalloys

Teledyne Allvac

P.O. Box 5030
Monroe, NC 28110-4018
USA

phone: 800 537 5551
fax: 704 289 6133

Iron-base superalloys; Nickel, alloys, superalloys; Titanium and alloys

Teledyne Cutting Tool

No. 1 Teledyne Place
La Vergne, TN 37086-1984
USA

phone: 615 641 4293
fax: 615 641 4333

Refractory metals; High alumina and "cermets";
Carbides; Nitrides; Carbon, graphite, diamond

Teledyne Pittsburgh Tool Steel

1535 Beaver Avenue
Monaca, PA 15061
USA

phone: 412 773 7000
fax: 412 774 4069

Tool steels

Teledyne Powder Alloys

350 Allwood Road
Clifton, NJ 07012
USA

phone: 201 773 4850

Refractory metals

Teledyne Rodney Metals

P.O. Box 6915
New Bedford, MA 02742-0915
USA

phone: 508 996 5691
fax: 508 993 3176

Stainless steels, sheet, strip; Copper and alloys;
Nickel, alloys, superalloys; Titanium and alloys

OTHER LOCATIONS

Teledyne Rodney Metals of California
7305 Paramount Blvd.
Pico Rivera, CA 90660
USA
phone: 800 423 5564
fax: 310 949 8640

Teledyne Rodney Metals of Germany
4322 Sprockhoevel 1
Kleinbeckstrasse 7 Postfach 132
Germany
phone: 49 2324 7532
fax: 49 2324 7536

Teledyne Rodney Metals of Carolina
1000 Ryan Avenue
Walterboro, SC 29488
USA
phone: 800 553 4351
fax: 803 538 5266

Teledyne Rodney Metals of Missouri
10855 Metro Court
Maryland Heights, MO 63043
USA
phone: 800 325 1455
fax: 314 993 8826

Teledyne Rodney Metals of France
Z1 Les chaumes
18570 La Chapelle Sa
France
phone: 33 4826 5354
fax: 33 4826 5108

Teledyne Rodney Metals of Pennsylvania
1045 Pulinski Road
Ivyland, PA 18974
USA
phone: 800 638 2571
fax: 215 675 7860

Teledyne Ryan Aeronautical

P.O. Box 85311
San Diego, CA 92186-5311
USA

phone: 619 291 7311

Stainless steels

OTHER LOCATION
Teledyne Ryan Aeronautical

2701 Harbor Drive
San Diego, CA 92112
USA

Teledyne Vasco

P.O. Box 151
Rt. 981 North
Latrobe, PA 15650-0151
USA

phone: 800 537 5551
fax: 412 537 3708

Carbon steels, mill shapes; Alloy steels, mill shapes; Tool steels

Teledyne Wah Chang Albany

P.O. Box 460
Albany, OR 97321-0136
USA

phone: 503 926 4211
fax: 503 976 6990

Refractory metals; Titanium and alloys; Vanadium; Zirconium

Telford Shell Mouldings Ltd.

Units D1/D2
Halesfield 21
Telford, Shropshire TF7 4NX
England

phone: 0952 580946
fax: 0952 582546

Copper and alloys; Tin and alloys

Tetra Plastics Inc.

620 Spirit of St.Louis Blvd.
Chesterfield, MO 63005
USA

phone: 314 532 3655
fax: 314 532 7981

Thermoplastic elastomers

Textron Inc.

CWC Castings Division
1085 W.Sherman Blvd.
Muskegon, MI 49441-3588
USA

phone: 616 733 1331
fax: 616 739 2649

Cast iron

Textron Specialty Materials

Subsidiary of Textron Inc.
2 Industrial Avenue
Lowell, MA 01851
USA

phone: 508 452 8961
fax: 508 454 5619

Acrylics; Epoxies; Polyurethanes; Carbon, graphite, diamond; Fibers

The Andersons

PO Box 119
1200 Dussel Dr.
Maumee, OH 43537
USA

phone: 800 472 3220
fax: 419 891 6659

Fillers, additives

The Esab Group Inc.

Welding Consumables HDQ
P.O. Box 517
Hanover, PA 17331
USA

phone: 717 637 8911
fax: 717 637 7651

Ferrous welding, hard facing electrodes

The Miller Company

99 Center Street
Meriden, CT 06450
USA

phone: 203 235 4474
fax: 203 634 6926

Copper and alloys

The Polymer Corp.

2120 Fairmont Ave.
Reading, PA 19612-4235
USA

phone: 800 433 5189
fax: 800 366 0801

Acetals; Fluoropolymers, chloropolymers; Polyamides; Polyamides; Polyamide-imides; Polyesters, thermoplastic; Polyesters, thermoset; Polystyrenes

Thermofil Inc.

PO Box 489
6150 Whitmore Lake Rd.
Brighton, MI 48116-0489
USA

phone: 800 444 4408
fax: 313 227 3824

Acetals; Polyamides; Polycarbonates; Polyesters, thermoplastic; Polyketones; Polyisoimides; Polyolefins; Polyphenylenes; Polyurethanes; Polystyrenes; Polysulfones; Polyvinyl chlorides

Thermoset Plastics Inc.

5101 East 65th St.
PO Box 20902
Indianapolis, IN 46220-0902
USA

phone: 317 259 4161
fax: 317 252 8402

Epoxies; Polyurethanes; Silicones

Thomas & Skinner Inc.

P.O. Box 150-B
Indianapolis, IN 46206
USA

phone: 317 923 2501
fax: 317 923 5919

Ferrous electronic materials

Thomas Steel Strip Co.

Delaware Ave. N.W.
Warren, OH 44485-2699
USA

phone: 800 321 7778
fax: 216 841 6260

Carbon steels, sheet, strip

Thomson Industries Inc.

2 Shore Road
Port Washington, NY 11050
USA

phone: 800 645 9357

Tool steels

Three M Company

Aerospace Materials Dept.
Bldg. 223-1N-07
3M Center
St. Paul, MN 55144-1000
USA

phone: 800 235 2376
fax: 612 733 7416

Epoxies

Thyssen Edelstahlwerke Inc.

Welding Products Div.
365 Village Dr.
Carol Stream, IL 60188
USA

phone: 800 323 1233

Alloy steels, mill shapes; Alloy steels, sheet, strip; Alloy steels, plate; Stainless steels, mill shapes; Stainless steels, sheet, strip; Ferrous welding, hard facing electrodes; Tool steels

TIME GmbH

Eschborner Landstr, 164
6000 Frankfurt 90
Germany

phone: 49 69 78 3011
fax: 49 69 78 3013

Nickel, alloys, superalloys

Timken Co.

1835 Dueber Avenue S.W.
GNW-19
Canton, OH 44706-2798
USA

phone: 216 471 4154

Alloy steels, mill shapes; Stainless steels, mill shapes; Tool steels

Titanium Metals Corp.

1999 Broadway
43rd Floor
Denver, CO 80202
USA

phone: 303 296 5600

Titanium and alloys

Toshiba Ceramics Co., Ltd.

Shinjuku Nomura Bldg.,26-2,
Nishi-Shinjuku 1-chome
Shinjuku,Tokyo 163
Japan

phone: 81 3 348 7411
fax: 81 3 343 8627

Electrical ceramics; Refractory ceramics; Engineering ceramics; Nitrides; Glass, quartz; Carbon, graphite, diamond; Oxides

Toshiba Tungaloy Co., Ltd.

755 Greenleaf Avenue
Elk Grove Village, IL 60007
USA

phone: 708 806 8800

High alumina and "cermets"

TradeARBED Inc.

825 Third Ave.
New York, NY 10022
USA

phone: 212 486 9890
fax: 212 355 2159

Carbon steels, mill shapes

Trent Tube

Div. of Crucible Materials Cor
2015 Energy Drive
P.O. Box 77
East Troy, WI 53120
USA

phone: 414 642 7321

Stainless steels, mill shapes

Tri-Clover

9201 Wilmot Rd.
Kenosha, WI 53141
USA

phone: 414 694 5511

Cast iron; Nickel, alloys, superalloys

TSE Industries Inc.

5260 113th Ave.North
Clearwater, FL 34620
USA

phone: 800 237 7634
fax: 813 527 0487

Acetals; Acrylics; Epoxies; Phenolics; Polyamides; Polycarbonates; Polyesters, thermoplastic; Polyketones; Polyolefins; Polyurethanes; Polystyrenes

Tungum Hydraulics Ltd.

The White House, Arle
Cheltenham
Gloucestershire Gl 51 0AD
England

phone: 44 0242 245521
fax: 44 0242 222 656

Copper and alloys; Zinc and alloys

Tyler, W.S., Inc.

Combustion Engineering Inc.
P. O. Box 8900
Gastonia, NC 28053
USA

phone: 800 238 9537

Alloy steels

OTHER LOCATION

W.S. Tyler Inc.

Combustion Engineering Inc.
8570 Tyler Blvd.
Mentor, OH
USA

U.N. Alloy Steel Corp.

300 Ocean Avenue
Revere, MA 02151
USA

phone: 617 284 6300

Tool steels

U.S. Reduction Co. (USCO)

9200 Calumet Ave.
Munster, IN 46321
USA

phone: 219 836 0555/51
fax: 219 836 0058

Aluminum and alloys

U.S. Steel Group

a Unit of USX Corporation
600 Grant St.
Room 1844
Pittsburgh, PA 15219-2749
USA

phone: 412 433 3974

Carbon steels, sheet, strip; Carbon steels, plate

Ube Industries Ltd

Ube Bldg, 3-11, Higashi-shinag
Shinagawa-ku
Tokyo 140
Japan

phone: 81 03 5460 3326
fax: 81 03 5460 3420

Polyamides

Uddeholm Corp.

4902 Tollview Drive
Rolling Meadows, IL 60008
USA

phone: 708 577 2220
fax: 708 577 8028

Stainless steels; Tool steels

OTHER LOCATIONS

Aceros ASSAB Mexico SA de C.V.
Calle Ocho No. 2 Letra"C"
Fracc Ind. Alco Blanco
53370 Naucalpan De Juarez
Estado De Mexico
Mexico
phone: 011 525 576 54
fax: 011 525 576 68

Aceros ASSAB Mexico SA de C.V.
Lerdo De Tejada No. 542
San Nicolas De Los Garza
Monterrey, Nuevo Leon
Mexico
phone: 011 52 83 52 52
fax: 011 52 83 52 53

Uddeholm Corp.
7901 Sovereign Row
Dallas, TX 75247
USA
phone: 214 630 3400
fax: 214 631 1358

Uddeholm Corp.
6396 Product Drive
Sterling Heights, MI 48312
USA
phone: 313 939 0100
fax: 313 939 1132

(continued)

Uddeholm Corp. (cont'd)

Uddeholm Corp.
851 N. Expressway Drive
Itasca, IL 60143
USA
phone: 708 773 2525
fax: 708 773 4114

Uddeholm Corp.
9331 Santa Fe Springs Rd.
Santa Fe Springs, CA 90670
USA
phone: 310 946 6503
fax: 310 946 4274

Uddeholm Corp.
5701 Mableton Parkway
Suite 4-D
Mableton, GA 30059
USA
phone: 404 944 3200
fax: 404 944 0580

Uddeholm Corp.
23529 Connecticut St.
Hayward, CA 94545
USA
phone: 510 782 5211
fax: 510 887 7468

Uddeholm Corp.
40 Robbie Road
Avon, MA 02322
USA
phone: 508 587 7708
fax: 508 587 7556

Uddeholm Corp.
4540 E. 71st Street
Cleveland, OH 44105
USA
phone: 800 321 3014
fax: 216 271 0064

Uddeholm Corp.
7900 Hub Parkway
Cleveland, OH 44125
USA
phone: 216 524 8770
fax: 216 642 1042

Uddeholm Corp.
16140 S.W. 72nd Avenue
Portland, OR 97224
USA
phone: 503 620 1032
fax: 503 620 3294

Uddeholm Corp.
3991 S. Mariposa Street
Englewood, CO 80110
USA
phone: 303 761 7762
fax: 303 781 9021

Uddeholm Ltd.
6659 Ordan Drive
Mississauga, Ontario L5T 1K6
Canada
phone: 416 670 8333
fax: 416 673 8249

(continued)

Uddeholm Corp. (cont'd)

Uddeholm Ltd.
Annacis Business Park
730 Eaton Way, Unit 10
New Westminster, BC V3M 6J9
Canada
phone: 604 525 3354
fax: 604 525 7013

Uddeholm Ltd.
3521 Rue Ashby
St. Laurent, Quebec H4R 2K3
Canada
phone: 514 333 8000
fax: 514 331 0542

Ulbrich Stainless Steels & Special Metal

57 Dodge Ave.
North Haven, CT 06473
USA
phone: 203 239 4481
fax: 203 239 7479

Stainless steels, sheet, strip; Ferrous electronic materials; Cobalt, alloys, superalloys; Nickel, alloys, superalloys; Titanium and alloys

Uniform Tubes Inc.

200 W. 7th Ave.
Collegeville, PA 19426
USA
phone: 215 539 0700
fax: 215 489 1150

Carbon steels, mill shapes; Stainless steels, mill shapes; Cast iron; Aluminum and alloys; Copper and alloys; Precious metals; Refractory metals; Titanium and alloys; Zirconium

(continued)

Uniform Tubes Inc. (cont'd)

OTHER LOCATIONS

UTI Corporation
Suite 235
16801 Addison Rd.
Dallas, TX 75248
USA

UTI Corporation
Suite 325, Third Floor
1501 Westvliff Drive
Newport Beach, CA 92660
USA

UTI Corporation
Suite 203
414 Plaza Drive
Westmont, IL 60559
USA

UTI, UK, Ltd.
P.O. Box 2
Shanklin, Isle of Wight PO 37 6UU
U.K.

Union Carbide Corp.

Umetco
2754 Compass Drive
Suite 280
Grand Junction, CO 81506
USA

phone: 303 245 3700
fax: 303 245 7543

Refractory metals; Titanium and alloys; Acetals; Acrylics; Cellulosics; Epoxies; Phenolics; Polyesters, thermoplastic; Polyolefins; Polyvinyl chlorides; Thermoplastic elastomers; Carbides; Nitrides; Glass, quartz; Carbon, graphite, diamond; Fibers

OTHER LOCATIONS

Union Carbide Corp.
Specialty Polymers & Composite
39 Old Ridgebury Rd.
Danbury, CT 06817-0001
USA
phone: 203 794 5300

UCAR Carbon Company Inc.
P.O. Box 94617
Parma, OH 44130
USA
phone: 800 342 3698

(continued)

Union Carbide Corp. (cont'd)

Union Carbide Coatings Service Corp.
1500 Polco St.
Indianapolis, IN 46224
USA
phone: 800 822 7284

Union Carbide Corp.
UCAR Emulsion Systems
410 Gregson Drive
Cary, NC 27511
USA
phone: 919 469 6700

Union Carbide Corp.
International Petrochemicals D
39 Old Ridgebury Rd.
Danbury, CT 06817
USA
phone: 203 794 2000

Union Carbide Corp.
UCAR Coatings Resins Departmen
200 Cottontail Lane (Bldg. 5)
P.O. Box 6783
Somerset, NJ 08873
USA
phone: 800 223 0537

Union Carbide Corp.
Polyolefins Division
2001 Ford Circle
Milford, OH 45150
USA
phone: 513 248 0303

Union Carbide Corp.
Organosilicon Products
P.O. Box 38002
S. Charleston, WV 25303-3802
USA
phone: 800 523 5862

Union Carbide Corp.
Solvents & Coatings Materials Div.
10 South Riverside Plaza
Chicago, IL 60606
phone: 312 454 2000

Union Carbide Corp.
Advanced Ceramics
P.O. Box 94924
Cleveland, OH 44101
USA
phone: 800 822 4322

Uniroyal Chemical Company, Inc.

World Headquarters
Benson Road, 2-31
Middlebury, CT 06749
USA

phone: 800 243 2650

Polyurethanes; Thermoplastic elastomers

OTHER LOCATIONS

Premier Chemical Ltd.
Bank Tower, Suite 704
205 Tun Hwa North Road
Taipei
Taiwan, R.O.C.
phone: 011 866 2 713 1
fax: 011 886 2 713 3

Uniroyal Chemical Ltd.
P.O. Box 160
Erb Street
Elmira, Ontario N3B 3A3
Canada
phone: 519 669 1671
fax: 519 669 1679

Uniroyal Chemical Asia, Ltd.
51 Newton Road, #14-04/05
Goldhill Plaza
Singapore 1130
Singapore
phone: 011 65 352 1411
fax: 011 65 352 2397

Uniroyal Chemical Ltd.
Kennet House
4 Langley Quay
Waterside Drive
Slough, Berkshire SL3 6EH
England
phone: 011 44 753 5808
fax: 011 44 753 5913

Uniroyal Chemical Company, Inc.
Douglas Center
Suite 611
2600 Douglas Road
Coral Gables, FL 33134
USA
phone: 305 446 1856
fax: 305 443 0759

Uniroyal Chemical Pty. Ltd.
Unit 2
13 Stanton Road
Seven Hills, NSW 2147
Australia
phone: 011 61 2 838 78
fax: 011 61 2 838 78

(continued)

Uniroyal Chemical Company, Inc. (cont'd)

Uniroyal Chimica S.r.l.
Viale delle Industrie, 40
Casella Postale 13
04013 Latina Scalo (
Italy
phone: 011 39 773 6301
fax: 011 39 773 6310

Uniroyal Quimica, S.A.
Avenida Morumbi, 7.029
CEP 05650
Sao Paulo, SP 01000
Brazil
phone: 011 55 11 844 5
fax: 011 55 11 844 5

United American Metals Corp.

2246 W. Hubbard St.
Chicago, IL 60612
USA

phone: 312 733 6700

Tin and alloys

United States Bronze Powders Inc.

P.O. Box 31
Flemington, NJ 08822
USA

phone: 908 782 5454

Copper and alloys

United States Pipe & Foundry Co.

PO Box 10406
Birmingham, AL 35202
USA

phone: 205 254 7000
fax: 205 254 7494

Stainless steels, mill shapes

OTHER LOCATIONS

United States Pipe & Foundry Co.
Eastern Regional Office
Burlington, NJ 08016
USA
phone: 609 387 6000
fax: 609 387 6050

United States Pipe & Foundry Co.
Southwestern Regional Office
P.O. Box 260848
Plano, TX 75026-0848
USA
phone: 214 423 3881
fax: 214 423 4573

United States Pipe & Foundry Co.
Western Regional Office
P.O. Box 247
Westmont, IL 60559
USA
phone: 708 920 0050
fax: 708 920 0418

United States Pipe & Foundry Co.
Pacific Coast Regional Office
P.O. Box 827
Union City, CA 94587
USA
phone: 510 537 6690
fax: 510 471 1155

Unna Messingwerk, A.G.

Uelzener Weg 36
Postfach 1729

Unna, Westphalia D4750
Germany

Aluminum and alloys

Usinor Sacilor

4 Place de la Pyramide
Cedex 33
F-92070 Paris-La Def
France

phone: 33 1 49 00 60 2
fax: 33 1 49 00 60 4

Stainless steels, sheet, strip

UTP. Welding Materials, Inc.

P.O. Box 721678
Houston, TX 77272-1678
USA

phone: 713 499 1212
fax: 713 499 4347

Copper and alloys; Nickel, alloys, superalloys

Vacuumschmelze GmbH

c/o Siemens Components, Inc.
186 Wood Avenue South
Iselin, NJ 08830
USA

phone: 908 494 3530
fax: 908 603 5994

Stainless steels; Iron-base superalloys; Ferrous electronic materials; Cobalt, alloys, superalloys; Copper and alloys; Nickel, alloys, superalloys

OTHER LOCATION
Vacuumschmelze GmbH
V-L
37 Gruener Weg
Postfach 22 53
D-6450 Hanau 1
Germany

phone: 49 6181 38026
fax: 49 6181 3845

Valenite Inc.

P.O. Box 3950
750 Stephenson Hwy.
Troy, MI 48007-3950
USA

phone: 800 521 3878
fax: 313 583 2515

High alumina and "cermets"; Carbides; Nitrides; Carbon, graphite, diamond

OTHER LOCATIONS

Barth-Werkzeugmaschinen
Industriestrasse
A-2345 Brunn bei Wie
Austria
phone: 43 022 36 31 66
fax: 43 022 36 3 16

Buck & Hickman Ltd.
Bank House
100 Queen Street
Sheffield S1 2DW
England
phone: 44 0742 76 66 6
fax: 44 0742 76 69 5

Industrial Tools S.C. Sorensen
Tronholmen 14
8900 Randers
Denmark
phone: 45 6 42 11 11
fax: 45 6 41 82 66

John Rainey & Son
219 Hillhall Road
Lisburn BT 27 5JA
Ireland
phone: 353 0846 67 531
fax: 353 0846 67 15

Nippon Valenite-Modco K.K.
21-10 Kita-Kohjiya 1-chome
Ohta-Ku
Tokyo 144
Japan
phone: 81 3 743 5711
fax: 81 3 743 5710

Tool Center A.B.
P.O. Box 4019
Hantverkarvagen 12
18304 Taby
Sweden
phone: 46 8 756 70 70
fax: 46 8 756 4810

V.M.I. Tooling Comp. (Pty.) Ltd.
24-26 Patrick Road
Jet Park Boksburg
1620 Tvl. S.A.
South Africa
phone: 27 11 826 1541
fax: 27 11 826 1114

Valenite Inc.
Walmet Division
404 E. 10 Mile Road
Pleasant Ridge, MI 48069
USA
phone: 313 589 6400
fax: 313 546 6066

(continued)

Valenite Inc. (cont'd)

Valenite Inc.
Poligono Industrial
28820 Coslada
Madrid
Spain
phone: 34 6 72 81 48
fax: 34 6 72 9860

Valenite-Modco (Australia)
7 Acheson Place
North Coburg, Victoria 3058
Australia
phone: 61 3 3 50 48 33
fax: 61 3 350 40 53

Valenite-Modco (U.K.) Ltd.
Unit 12, Alston Drive
Bradwell Abbey Industrial Esta
Milton Keynes MK 13 9HA
England
phone: 44 0908 31 98 8
fax: 44 0908 31 14 0

Valenite-Modco (U.K.) Ltd.
Clydeway Industrial Centre
Unit 5C
45 Finnieston Street
Glasgow G3 8JU
Scotland
phone: 44 41 204 1755
fax: 44 41 204 0773

Valenite-Modco GmbH
Postfach 1654
6920 Sinsheim
Germany
phone: 49 7261 6940
fax: 49 7261 648 01

Valenite-Modco GmbH
Murgenthalstr. 15
4900 Langenthal
Switzerland
phone: 41 063 23 25 75
fax: 41 063 22 89 30

Valenite-Modco Industria e Comercio Ltda.
Rua Braganca Paulista 1036
Santo Amaro CEP 04 727
Sao Paulo
Brazil
phone: 55 2 47 71 22
fax: 55 5 21 84 72

Valenite-Modco International, Inc.
Taiwan Branch
3rd Floor-4,35 Nan Hua
1st Street
Lu Chu, Tao-Yuan
Taiwan, R o. C
phone: 886 033 228432/
fax: 886 033 525028

(continued)

Valenite Inc. (cont'd)

Valenite-Modco Pacific Co. Ltd.
Sama Trading Co., Ltd.
11F. Suwoon Hoikwan
88. Kyungun-Dong, Jongno-Ku
Youngdong P.O. Box 512
Seoul
South Korea
phone: 82 8 227 34 063
fax: 82 8 227 34 064

Valenite-Modco Pte. Ltd.
No. 25 Pandan Loop, Block 12
Pandan Light Industries Park
Singapore 0512
Singapore
phone: 65 779 7455
fax: 65 778 0992

Valenite-Modco S.a.r.l.
1031 e. Ch.d'Alsemberg, b.124
B-1180 Bruxelles
Belgium
phone: 32 2 332 1071
fax: 32 2 332 0065

Valenite-Modco S.a.r.l.
B.P. 65
42162 Andrezieux-Bou
France
phone: 33 77 36 55 52
fax: 33 77 55 64 22

Valenite Mexicalli S.A. de C.V.
Av. Convento de Santa Monica N
Jardines de Santa Monica
54050 Tiainepantia d
Mexico
phone: 52 5 398 7878
fax: 52 5 398 7829

Valenite-Modco S.r.l.
Direzione Commerciale
Via C. Battisti 3
20070 Vizzolo Predabissi
Italy
phone: 39 02 983 0021
fax: 39 02 983 5607

Valenite-Modco Ltd.
Unit No. 1
1456 Wallace Road
Oakville, Ontario L6L 2Y2
Canada
phone: 416 827 6040
fax: 519 948 9691

Valimet, Inc.

PO Box 6186
431 Sperry Road
Stockton, CA 95206
USA

phone: 209 982 4870
fax: 209 982 1365

Aluminum and alloys

Vallourec S.A.

3 Post Oak Central
Suite 710
1990 Post Oak Blvd.
Houston, TX 77056
USA

phone: 713 961 2468
fax: 713 961 4316

Stainless steels, sheet, strip; Iron-base superalloys;
Nickel, alloys, superalloys; Titanium and alloys

OTHER LOCATIONS

Vallourec GmbH
Konigsallee 70
Postfach 200 808
D-4000 Dusseldorf 1
Germany
phone: 49 211 130 960
fax: 49 211 326 683

Vallourec Inc.
414 Allegheny River Boulevard
Suite 205
Oakmont, PA 15139
USA
phone: 412 828 7520
fax: 412 828 4386

Vallourec Industries
Industrial Applications Group
130 rue de Silly
BP 415-92103 Boulogn
France
phone: 33 1 49 09 35 0
fax: 33 1 49 09 37 0

Vallourec Middle East Delegation
P.O. Box 4015
Abu Dhabi
United Arab Emi
phone: 971 2 338 778

(continued)

Vallourec S.A. (cont'd)

Valtubes UK Ltd.
Henley Court
High Street
Henley in Arden
West Midlands B955BA
Great Britain

phone: 44 0564 792277
fax: 44 0564 795032

VAW Vereinigte Aluminium-Werke AG

Postfach 24 68
Georg-von-Boeselager-Strasse 2
D-5300 Bonn 1
Germany

phone: 49 0228/5 52 02
fax: 49 0228 5 52 22

Aluminum and alloys

Vereinigte Chemische Fabriken

Georg-von-Boeselager Str 25
Bonn 1 5300 0334
Germany

Refractory metals

Vereinigte Metallwerke Ranshofen-Berndor

A-5282 Braunau-Ransh
Austria

Copper and alloys

Viking Engineered Cast Products

A Unit of IDEX Corp.
715 Viking Road
P.O. Box 8
Cedar Falls, IA 50613-0008
USA

phone: 319 273 8170
fax: 319 273 8156

High alumina and "cermets"

Viskase Corp.

6855 W. 65th Street
Chicago, IL 60638
USA

phone: 708 496 4200

Polyvinyl chlorides

Voest-Alpine International, Inc.

60-A E. 42nd Street
New York, NY 10165
USA

phone: 212 661 1060

Carbon steels, plate

Wabash Alloys, Inc.

Div. of Connell Ltd. Partnersh
Old Hwy. 24 West
PO Box 466
Wabash, IN 46992
USA

phone: 219 563 7461

Aluminum and alloys

Wall Colmonoy Corp.

30261 Stephenson Highway
Madison Heights, MI 48071-1650
USA

phone: 313 585 6400
fax: 313 585 7960

Stainless steels; Ferroalloys, pure iron; Ferrous welding, hard facing electrodes; Cobalt, alloys, superalloys; Copper and alloys; Magnesium and alloys; Nickel, alloys, superalloys; Refractory metals

Washington Penn Plastic Co.,Inc.

2080 North Main St.
PO Box 236
Washington, PA 15301
USA

phone: 412 228 1260

Polyolefins; Polystyrenes

Washington Steel Corp.

subsidiary of Lukens Steel Co.
P.O. Box 494
Washington, PA 15301
USA

phone: 412 222 8000

Carbon steels; Alloy steels; Stainless steels

Waterbury Rolling Mills Inc.

240 E. Aurora Street
Waterbury, CT 06708
USA

phone: 203 754 0151

Copper and alloys

Watsontown Foundry

P.O. Box 126
Watsontown, PA 17777
USA

phone: 717 538 5546

Cast iron

Waukesha Foundry, Inc.

1300 Lincoln Ave.
Waukesha, WI 53186
USA

phone: 414 542 0741
fax: 414 549 8440

Copper and alloys; Nickel, alloys, superalloys

Weardale Steel (Wolsingham) Ltd.

Wolsingham, Bishop A, Co.
Durham DL13 3HX
England

phone: 44 0388 527 201
fax: 44 0388 527 838

Alloy steels; Stainless steels; Nickel, alloys, superalloys; Refractory metals; Carbides

Weatherly Casting & Machine Co.

300 Commerce Street
P.O. Box 21
Weatherly, PA 18255
USA

phone: 717 427 8611
fax: 717 427 9895

Manganese

Webco

Southwest Tube Division
P.O. Box 100
201 S. Woodland Drive
Sand Springs, OK 74063
USA

phone: 800 331 4096
fax: 918 245 0306

Carbon steels, mill shapes

OTHER LOCATION

Webco Industries, Inc.
Stainless Division
P.O. Box 100
201 South Woodland Dr.
Sand Springs, OK 74063
USA

phone: 800 331 4096
fax: 918 865 4996

Welded Tube Company of America

1855 E. 122nd St.
Chicago, IL 60633
USA

phone: 800 733 5683
fax: 312 646 6128

Carbon steels, mill shapes

Weldwire Co.,Inc.

103 Queens Drive
King of Prussia, PA 19406
USA

phone: 215 242 1100

Copper and alloys

Wellman Dynamics Corp.

U.S. Rte. 34
P.O. Box 147
Creston, IA 50801
USA

phone: 515 782 8521
fax: 515 782 4844

Aluminum and alloys

Wellman Inc.

Engineering Resins Division
PO Drawer 188
Highway 41/51 North
Johnsonville, SC 29555-0188
USA

phone: 803 386 2011
fax: 803 386 8273

Polyamides

Wellman, S.K., Corp.

6180 Cochran Rd.
Solon, OH 44139
USA

phone: 216 498 2275
fax: 216 498 2290

Carbon, graphite, diamond

WESGO

477 Harbor Blvd.
Belmont, CA 94002
USA

phone: 415 592 9440
fax: 415 598 3236

Copper and alloys; Precious metals; Engineering ceramics; High alumina and "cermets"; Nitrides

OTHER LOCATION

WESGO, Inc.

24 Madison Road
Fairfield, NJ 07006
USA

phone: 201 227 8877

Western Reserve Mfg. Co.

5311 West River Rd. N.
Lorain, OH 44055
USA

phone: 216 277 1226
fax: 216 277 9526

Copper and alloys

Westig

Provincial House
Solly Street
Sheffield S1 4BA
England

phone: 44 0742 739 544
fax: 44 0742 757 375

Alloy steels; Stainless steels; Tool steels

OTHER LOCATIONS

Westig GmbH & Co. KG
Postfach 2040
Hochstresse 32
D-4750 Unna
Germany
phone: 49 02303 2030
fax: 49 0 2303 203 2

Westig of America, Inc.
P.O. Box 839
26 Chapin Road
Pine Brook, NJ 07058
USA
phone: 201 882 9558
fax: 201 882 8091

Westinghouse Electric Corp.

Commercial Nuclear Fuel Divisi
1310 Beulah Road
P.O. Box 355
Pittsburgh, PA 15230
USA

phone: 412 374 2273

Refractory metals; Zirconium

Westlake Plastics Co.

PO Box 127
West Lenni Road
Lenni, PA 19052
USA

phone: 215 459 1000
fax: 215 459 1084

Polycarbonates; Polyolefins; Polystyrenes; Polysulfones

Wheeling-Pittsburgh Steel Corp.

1134 Market St.
Wheeling, WV 26003
USA

phone: 304 234 2400
fax: 304 234 2539

Carbon steels, sheet, strip

(continued)

Wheeling-Pittsburgh Steel Corp. (cont'd)

OTHER LOCATION

Wheeling-Pittsburgh Steel Corp.
4 Gateway Ct.
16th Floor
Pittsburgh, PA 15222
USA
phone: 412 288 3511
fax: 412 288 3555

Wieland-Werke AG Metallwerke

Technische Kundenberatung
Postfach 4240
Graf-Arco Strasse 38
D-7900 Ulm
Germany
phone: 49 07 31 9 440
fax: 49 07 31 9 2772

Copper and alloys

OTHER LOCATIONS

Wieland (Schweiz) AG
Eichstrasse 39
CH-6330 Cham
Switzerland
phone: 41 42 36 46 55
fax: 41 42 36 18 85

Wieland Italia S.R.L.
Via G Frua, 22
I-20146 Milano
Italy
phone: 39 2 48 00 01 8
fax: 39 2 48 00 81 5

Wieland Benelux S.A.
Rue Forestiere 22, B.6
B-1050 Brussel
Belgium
phone: 32 2 640 80 40
fax: 32 2 640 50 99

Wieland Metals Canada Ltd.
41 Lambeth Road
Islington, Ontario M9A 2Y6
Canada
phone: 416 232 0451
fax: 416 232 2650

(continued)

Wieland-Werke AG Metallwerke (cont'd)

Wieland Metals Service Center Inc.
567 Northgate Parkway
Wheeling, IL 60090
USA
phone: 708 537 3990
fax: 708 537 4085

Wieland S.A. (Pty.) Ltd.
P.O. Box 1048
Bedfordview 2008
South Africa
phone: 27 11 6098 32 5
fax: 27 11 452 37 91

Wieland Metals Singapore (Pte) Ltd.
10 Tuas Road
Singapore 2263
Singapore
phone: 65 861 9788
fax: 65 861 9800

Wieland S.A.R.L.
P.A. Parisest
Rue L. Jouhaux
Croissy-Beaubourg
F-77324 Marne L.V. C
France
phone: 33 1 601 71 92
fax: 33 1 601 717 08

Wilkinson Company

31011 Agoura Road
PO Box 4558
Westlake Village, CA 91359
USA

phone: 818 889 0050
fax: 818 889 3810

Copper and alloys; Nickel, alloys, superalloys; Precious metals

Williams Advanced Materials, Inc.

Subsidiary, Brush Wellman Inc.
2978 Main Street
Buffalo, NY 14214-1099
USA

phone: 716 837 1000
fax: 716 833 2926

Beryllium and alloys; Copper and alloys; Precious metals

OTHER LOCATION
Williams Advanced Materials (Far East Pte.) Ltd.
No. 10 Arumugam Road
Lion Industrial Bldg. #05-00
Singapore 1440
Singapore

phone: 011 65 741 9800
fax: 011 65 746 8433

Wolverine Tube Inc.

1010 Clarke Road
P.O. Box 6515, Station D
London, Ontario N5W 5S9
Canada

phone: 800 265 9271
fax: 519 455 9238

Copper and alloys

OTHER LOCATION
Wolverine Tube Inc.
P.O. Box 2202
2100 Market St. N.E.
Decatur, AL 35609-2202
USA

phone: 205 353 1310
fax: 205 355 6462

Worthington Steel & Annealing Co.

Div. of Worthington Industries
1205 Dearborn Dr.
Columbus, OH 43085
USA

phone: 614 438 3210
fax: 614 438 3171

Carbon steels

Xaloy Inc.

P.O. Box 1441
Pulaski, VA 24301
USA

phone: 703 980 7560
fax: 703 980 5670

Iron-base superalloys; Cast iron; Cobalt, alloys, superalloys; High alumina and "cermets"

Yoonsteel (Malaysia) Sdn. Bhd.

Site 14, Tasek Industrial Esta
Ipoh, Perak 31400
Malaysia

phone: 60 05 551 700
fax: 60 05 561 506

Carbon steels, mill shapes

Zapp, Robert, Werkstofftechnik GmbH

Goltsteinstr. 28
4000 Dusseldorf 1
Germany

phone: 211 35500
fax: 211 3550 200

Alloy steels, mill shapes; Stainless steels; Tool steels

Zeelan Industries Inc.

3M Center
Bldg. 220-8E-04
St. Paul, MN 55144
USA

phone: 612 737 1751
fax: 612 737 1764

Fillers, additives

Zircar Products, Inc.

110 North Main St.
Florida, NY 10921
USA

phone: 914 651 4481

Electrical ceramics; Refractory ceramics; Oxides

Zoltek

3101 McKelvey Road
St. Louis, MO 63044
USA

phone: 314 291 5110
fax: 314 291 8536

Fibers

Section II: Material Listing

Carbon steels

- Acciaierie Valbruna SpA
- Acenor-Vitoria
- Acos Villares Sede
- AFORA (Aceros Afora S.A.)
- Bethlehem Steel Corp.
- Cambridge Inc.
- Newport Steel Corp.
- Pechiney
- Pelton Casteel, Inc.
- Washington Steel Corp.
- Worthington Steel & Annealing Co.

Carbon steels, cast

- Eagle Alloy Inc.
- Empire Steel Castings Inc.
- Grede Foundries, Inc.
- Greenville Metals, Inc.
- Sivyer Steel Corp.
- SPS Technologies

Carbon steels, mill shapes

- AFORA (Aceros Afora S.A.)
- Atlantic Steel Co.
- Bethlehem Steel Corp.
- Bohler Edelstahl GmbH
- British Steel plc
- Bull Moose Tube Co.
- Bundy Corp.
- Calumet Steel Co.
- China Metallurgical Import & Export Corp.
- China Steel Corp.
- Daido Steel Co. Ltd.
- Diederichs, Karl Stahl-Walz und Hammerwerk
- Dofasco Inc.
- Ergst, Stahlwerke

Carbon steels, mill shapes (cont'd)

- Harsco Corp.
- Inland Steel Co.
- Kobe Steel USA Inc.
- LaSalle Steel Co.
- LTV Steel Co.
- Mukand Ltd.
- Ovako Steel
- Plymouth Tube Co., Extruded Products Div.
- Schwaebische Huettenwerke GmbH
- Seneca Wire & Mfg. Co.
- Sharon Tube Co.
- Steelmark-Eagle & Globe
- Sumitomo Metal America Inc.
- Tata Iron & Steel Co., Ltd.
- Teledyne Vasco
- TradeARBED Inc.
- Uniform Tubes Inc.
- Webco
- Welded Tube Co. of America
- Yoonsteel (Malaysia) Sdn. Bhd.

Carbon steels, sheet, strip

- Armco Steel Co.
- Bethlehem Steel Corp.
- British Steel plc
- China Metallurgical Import & Export Corp.
- China Steel Corp.
- Cold Metal Products Co., Inc.
- Daido Steel Co. Ltd.
- Empire-Detroit Steel Div.
- Ergst, Stahlwerke
- Follansbee Steel Corp.
- Gulf States Steel, Inc.
- Inland Steel Co.
- Kobe Steel USA Inc.
- LTV Steel Co.
- McLouth Steel Corp.
- Nippon Steel U.S.A. Inc.
- Rouge Steel Co.
- Samuel Whittar, Inc.
- Sidmar N.V.
- Steelmark-Eagle & Globe
- Sumitomo Metal America Inc.
- Thomas Steel Strip Co.
- U.S. Steel Group
- Wheeling-Pittsburgh Steel Corp.

Carbon steels, plate

- Acenor-Vitoria
- Armco Steel Co.
- Bethlehem Steel Corp.
- British Steel plc
- China Metallurgical Import & Export Corp.
- China Steel Corp.
- Gulf States Steel, Inc.
- Harsco Corp.
- Inland Steel Co.
- Kobe Steel USA Inc.
- Lukens Steel Co.
- Oregon Steel Mills Inc.
- Sumitomo Metal America Inc.
- U.S. Steel Group
- Voest-Alpine International, Inc.

Carbon steels, forged

- Bethlehem Steel Corp.

Carbon steels, powder

- Hoeganaes Corp.
- Powder Alloy Corp.

Alloy steels

- Acciaierie Valbruna SpA
- Acenor, S.A.
- Acenor-Vitoria
- Acme Steel Co.
- Acos Villares Sede
- AFORA (Aceros Afora S.A.)
- Albert Hoffmann GmbH
- Associated Spring Co.
- Aubert et Duval
- Bethlehem Steel Corp.
- CMW Inc.
- Cogne SpA
- Columbus McKinnon Corp.
- Continental Foundry & Machine Co.
- D-M-E Company
- Dorrenberg Edelstahl GmbH
- Duriron Co. Inc.
- Gustav Grimm Edelstahl-Werk GmbH
- Hamilton Technology, Inc.
- Kind & Co. Edelstahlwerk
- Krupp VDM
- Mitsubishi Materials Corp.
- Ossenberg & Cie, W.
- Pechiney
- Pelton Casteel, Inc.
- Scientific Alloys Inc.
- Sifbronze
- Tyler, W.S., Inc.
- Washington Steel Corp.
- Weardale Steel (Wolsingham) Ltd.
- Westig

Alloy steels, cast

- Capitol Castings Inc.
- Climax Specialty Metals
- Duraloy Blaw-Knox Inc.
- Eagle Alloy Inc.
- Empire Steel Castings Inc.
- ESCO Corp.
- Falk Corp., The
- Federal-Mogul Corp.
- Grede Foundries, Inc.
- Harrison Steel Castings Co.
- National Castings, Inc.
- Schmidt & Clemens GmbH & Co.
- Sivyer Steel Corp.

Alloy steels, mill shapes

- AFORA (Aceros Afora S.A.)
- AL Tech Specialty Steel Corp.
- Ambo-Stahl Gesellschaft
- Atlantic Steel Co.
- Aubert et Duval
- Bethlehem Steel Corp.
- BGH Edelstahl GmbH
- Bliss & Laughlin Steel Co.
- Bohler Edelstahl GmbH
- British Rolling Mills Ltd.
- British Steel plc
- Calumet Steel Co.
- Carpenter Technology Corp.
- China Metallurgical Import & Export Corp.
- Crucible Materials Corp.
- Daido Steel Co. Ltd.
- Deutsche Nickel AG
- Diederichs, Karl Stahl-Walz und Hammerwerk
- Dofasco Inc.
- Faitout Iron & Steel Co.
- Harsco Corp.
- Inland Steel Co.
- International Nickel Inc.
- Jessop Steel Co.
- Johnstown Corp.
- Kobe Steel USA Inc.
- Krupp Stahl AG
- Ladish Co., Inc.
- LaSalle Steel Co.
- Latrobe Steel Co.
- Lesjofors Steel AB
- LTV Steel Co.
- Mukand Ltd.
- Nachi Fujikoshi Corp.
- National Forge Co.
- Nippon Steel U.S.A. Inc.
- Ovako Steel
- Plymouth Tube Co., Extruded Products Div.
- Pyramid Steel Co.
- Quanex Corp.
- S.K.F. Industries Inc.
- Sanderson Kayser Ltd.
- Sanyo Special Steel Co., Ltd.
- Schmidt & Clemens GmbH & Co.
- Stahlwerk Stahlschmidt GmbH & Co.
- Stahlwerke R. & H. Plate
- Steelmark-Eagle & Globe
- Stelco Steel
- Sumitomo Metal America Inc.
- Tata Iron & Steel Co., Ltd.
- Teledyne Vasco
- Thyssen Edelstahlwerke Inc.
- Timken Co.
- Zapp, Robert, Werkstofftechnik GmbH

Alloy steels, sheet, strip

- Algoma Steel Corp. Ltd.
- American Tank & Fabricating Co.
- Bethlehem Steel Corp.
- British Steel plc
- China Steel Corp.
- Cold Metal Products Co., Inc.
- Daido Steel Co. Ltd.
- Deutsche Nickel AG
- Gulf States Steel, Inc.
- Inland Steel Co.
- Japan Steel Works Ltd.
- Kobe Steel USA Inc.
- Krupp Stahl AG
- Krupp-Hoesch-Stahl AG
- LTV Steel Co.
- McLouth Steel Corp.
- Middelburg Steel and Alloys (Pty) Ltd.
- Nippon Steel U.S.A. Inc.
- NKK Corp.
- Rouge Steel Co.
- Sidmar N.V.
- Steelmark-Eagle & Globe
- Stelco Steel
- Sumitomo Metal America Inc.
- Thyssen Edelstahlwerke Inc.

Alloy steels, plate

- Atlantic Steel Co.
- Bethlehem Steel Corp.
- British Steel plc
- China Steel Corp.
- Gulf States Steel, Inc.
- Harsco Corp.
- Inland Steel Co.
- International Nickel Inc.
- Japan Steel Works Ltd.
- Kobe Steel USA Inc.
- Lukens Steel Co.
- McLouth Steel Corp.
- Middelburg Steel and Alloys (Pty) Ltd.
- Nippon Steel U.S.A. Inc.
- NKK Corp.
- Oregon Steel Mills Inc.
- Stelco Steel
- Sumitomo Metal America Inc.
- Thyssen Edelstahlwerke Inc.

Alloy steels, forged

- Bethlehem Steel Corp.
- BGH Edelstahl GmbH

Alloy steels, powder

- Hoeganaes Corp.

Stainless steels

- Acciaierie Valbruna SpA
- Acenor, S.A.
- Acieries de Bonpertuis
- Acos Villares Sede
- AFORA (Aceros Afora S.A.)
- Albert Hoffmann GmbH
- Allegheny Ludlum Corp.
- Alloy Engineering & Casting Co.
- American Chain & Cable
- American Saw & Mfg. Co.
- Ancast, Inc.
- Anor, Acieries et Forges
- Apollo Metals Ltd.
- Armco Inc.
- Associated Spring Co.
- Aubert et Duval
- Babcock & Wilcox
- Beaumont Birch Co.
- Bethlehem Steel Corp.
- Blackalloy Co. of America, Inc.
- Bohler Edelstahl GmbH
- British Steel plc
- Cambridge Inc.
- Cannon-Muskegon Corp.
- Carpenter Technology Corp.
- Crane, Torry S., Co.
- Crucible Materials Corp.
- Cytemp Specialty Steel
- D-M-E Company
- Deloro Stellite Inc.

Stainless steels (cont'd)

- Dorrenberg Edelstahl GmbH
- Dupont, E.I., de Nemours & Co.
- Enpar Sonderwerkstoffe GmbH
- Ergst, Stahlwerke
- Gustav Grimm Edelstahl-Werk GmbH
- Harris, J. W., Co., Inc.
- Haynes International, Inc.
- Inco Alloys International, Inc
- Industrial Steels Inc.
- Johnson Mfg. Co.
- Junker, Otto, GmbH
- Kaydon Ring & Seal Inc.
- Kind & Co Edelstahlwerk
- Klockner-Werke AG
- Krupp VDM
- Metco Perkin-Elmer
- Ossenberg & Cie, W.
- Otis Elevator Co.
- Outokumpu
- Pechiney
- Precision Rolled Products, Inc.
- Remystahl
- Scientific Alloys Inc.
- Sifbronze
- Technetics Corp.
- Teledyne Ryan Aeronautical
- Uddeholm Corp.
- Vacuumschmelze GmbH
- Wall Colmonoy Corp.
- Washington Steel Corp.
- Weardale Steel (Wolsingham) Ltd.
- Westig
- Zapp, Robert, Werkstofftechnik GmbH

Stainless steels, cast

- Albert Hoffmann GmbH
- Alloy Engineering & Casting Co.
- ARC Mondial GmbH
- Atlas Foundry & Machine Co.
- Blackalloy Co. of America, Inc.
- Capitol Castings Inc.
- Carondelet Corp.
- Certified Alloy Products, Inc.
- Climax Specialty Metals
- Crane, Torry S., Co.
- Deloro Stellite Inc.
- Duraloy Blaw-Knox Inc.
- Eagle Alloy Inc.
- Empire Steel Castings Inc.
- ESCO Corp.
- Ferry-Capitain, Usines De Bussy
- Follsain Wearparts Foundry
- Greenville Metals, Inc.
- International Nickel Inc.
- Manoir Electro Alloys

Stainless steels, cast (cont'd)

- Nippon Stainless Steel Co. Ltd.
- Pose-Marre Edelstahlwerk GmbH
- Schmidt & Clemens GmbH & Co.
- Shellcast Corp.
- Sivyer Steel Corp.
- Stainless Foundry & Engineering
- Techni-Cast Corp.

Stainless steels, mill shapes

- Acme Tube
- AFORA (Aceros Afora S.A.)
- AL Tech Specialty Steel Corp.
- Ambo-Stahl Gesellschaft
- Avesta Sheffield Group
- BGH Edelstahl GmbH
- Bohler Edelstahl GmbH
- British Steel plc
- Carlson, G.O., Inc.
- Carpenter Technology Corp.
- Coshocton Stainless Div.
- Deutsche Nickel AG
- Duraloy Blaw-Knox Inc.
- Eastern Stainless Corp.
- Ergst, Stahlwerke
- Firth-Vickers Special Steels Ltd.
- Fisher Scientific Co.
- Greenville Tube Corp.
- Jessop Steel Co.
- Johnstown Corp.
- Krupp Stahl AG
- Krupp VDM
- Latrobe Steel Co.
- Lunkenheimer Co.
- Mukand Ltd.
- Nippon Stainless Steel Co. Ltd.
- Outokumpu
- Ovako Steel
- Plymouth Tube Co., Extruded Products Div.
- Sanderson Kayser Ltd.
- Sandvik Steel Co.
- Sanyo Special Steel Co., Ltd.
- Schmidt & Clemens GmbH & Co.
- Sheffield Forgemasters Ltd.
- Slater Steel Corp.
- Stahlwerk Stahlschmidt GmbH & Co.
- Stahlwerke R. & H. Plate
- Steelmark-Eagle & Globe
- Sumitomo Metal America Inc.
- SWEPCO Tube Corp.
- Thyssen Edelstahlwerke Inc.
- Timken Co.
- Trent Tube
- Uniform Tubes Inc.
- United States Pipe & Foundry Co.

Stainless steels, sheet, strip

- Acerinox, S.A.
- Allegheny Ludlum Corp.
- Armco Inc.
- Avesta Sheffield Group
- British Steel plc
- Carpenter Technology Corp.
- Coshocton Stainless Div.
- Deutsche Nickel AG
- Eastern Stainless Corp.
- Ergst, Stahlwerke
- Krupp Stahl AG
- Krupp VDM
- McLouth Steel Corp.
- Middelburg Steel and Alloys (Pty) Ltd.
- Nippon Stainless Steel Co. Ltd.
- NKK Corp.
- Osram Sylvania Inc.
- Outokumpu
- Sandvik Steel Co.
- Sheffield Forgemasters Ltd.
- Stahlwerk Stahlschmidt GmbH & Co.
- Steelmark-Eagle & Globe
- Teledyne Rodney Metals
- Thyssen Edelstahlwerke Inc.
- Ulbrich Stainless Steels & Special Metals, Inc.
- Usinor Sacilor
- Vallourec S.A.

Stainless steels, plate

- Allegheny Ludlum Corp.
- Avesta Sheffield Group
- British Steel plc
- Carlson, G.O., Inc.
- Eastern Stainless Corp.
- Krupp Stahl AG
- Middelburg Steel and Alloys (Pty) Ltd.
- NKK Corp.
- Sumitomo Metal America Inc.

Stainless steels, forged

- BGH Edelstahl GmbH
- Krupp VDM

Stainless steels, powder

- Hoeganaes Corp.
- MG Industries
- Powder Alloy Corp.
- Sanyo Special Steel Co., Ltd.

Ferroalloys, pure iron

- Billiton International Metals BV
- Globe Metallurgical Sales, Inc.
- Hoeganaes Corp.
- MicroMet Technology Inc.
- Pechiney Electrometallurgie
- Pyron Corp.
- Wall Colmonoy Corp.

Iron-base superalloys

- Allegheny Ludlum Corp.
- Aubert et Duval
- BGH Edelstahl GmbH
- British Steel plc
- Cannon-Muskegon Corp.
- Carboloy Inc.
- Carpenter Technology Corp.
- Dewramet Ltd.
- Duriron Co. Inc.
- Harrison Alloys Inc.
- Haynes International, Inc.
- Inco Alloys International, Inc
- International Nickel Inc.
- Krupp VDM
- PM Hochtemperatur-Metall GmbH
- Sheffield Forgemasters Ltd.
- Slater Steel Corp.
- Special Metals Corp.
- Steelmark-Eagle & Globe
- Teledyne Allvac
- Vacuumschmelze GmbH
- Vallourec S.A.
- Xaloy Inc.

Cast iron

- Advance Foundry Co.
- Anor, Acieries et Forges
- Blaw Knox Rolls
- Brillion Iron Works
- Burndy Corp.
- Castings USA
- Chambersburg Engineering Co.
- China Metallurgical Import & Export Corp.
- Climax Specialty Metals
- Continental Foundry & Machine Co.
- Crane Valves
- Crane, Torry S., Co.
- Deloro Stellite Inc.
- Denver Thomas, Inc.
- Duraloy Blaw-Knox Inc.
- Follsain Wearparts Foundry
- Georgia Iron Works Co.
- Grede Foundries, Inc.
- Gunite Corp.
- Hamilton Foundry
- Harrison Steel Castings Co.
- International Nickel Inc.
- Kaydon Ring & Seal Inc.
- Kelly Foundry & Machine Co.
- Macaulay Foundry Inc.
- Motor Castings Co.
- Nuclear Metals, Inc. (NMI)
- Otis Elevator Co.
- Peckovers, Ltd.

Cast iron (cont'd)

- QIT-Fer et Titane Inc.
- Sandusky International
- Scientific Alloys Inc.
- Sterling International Technology Ltd.
- Textron Inc.
- Tri-Clover
- Uniform Tubes Inc.
- Watsontown Foundry
- Xaloy Inc.

Tool steels

- Acciaierie Valbruna SpA
- Acenor, S.A.
- Acenor-Vitoria
- Acos Villares Sede
- AFORA (Aceros Afora S.A.)
- AL Tech Specialty Steel Corp.
- Albert Hoffmann GmbH
- Alloy Technology International, Inc.
- Ambo-Stahl Gesellschaft
- American Chain & Cable
- Anor, Acieries et Forges
- Associated Steel Corp.
- Aubert et Duval
- Babcock & Wilcox
- Barworth Flockton Ltd.
- Bendix Steel Co.
- Bethlehem Steel Corp.
- Billiton International Metals BV
- Black Drill Co., Inc.
- Black-Clawson Co.
- Bohler Edelstahl GmbH
- Brighton Electric Steel Casting Co.
- British Rolling Mills Ltd.
- British Steel plc
- Buffalo Wire Works Co.
- Burndy Corp.
- Carpenter Technology Corp.
- Cerro Metal Products Co.
- Champion Steel Co.
- China Metallurgical Import & Export Corp.
- Cook, George H. & Co. Ltd.
- Delsteel Inc.
- DoAll Co.
- Dorrenberg Edelstahl GmbH
- Erasteel Champagnole
- Faitout Iron & Steel Co.
- Finkl, A., & Sons Co.
- Firth-Vickers Special Steels Ltd.
- Great Western Steel Co.
- Greenville Metals, Inc.
- Gustav Grimm Edelstahl-Werk GmbH
- Hammond & Irving Inc.
- Jessop Steel Co.

Tool steels (cont'd)

- Johnstown Corp.
- Kind & Co Edelstahlwerk
- Kobe Steel USA Inc.
- Krupp Stahl AG
- Ladish Co., Inc.
- Latrobe Steel Co.
- Lesjofors Steel AB
- Lohmann, Friedr., GmbH
- Ludlow Steel Corp.
- Lukens Steel Co.
- Meridian Steel Co., Inc.
- Milne, A., & Co.
- Ossenberg & Cie, W.
- Otis Elevator Co.
- Paragon Steel & Tool Co.
- Peninsular Steel Co.
- Pennsylvania Steel Corp.
- Pettibone Corp.
- Precision Industries, Inc.
- Precision-Kidd Steel Co.
- Pyramid Steel Co.
- Remystahl
- Rex Buckeye Co. Inc.
- S.K.F. Industries Inc.
- Saarstahl AG
- Sanderson Kayser Ltd.
- Sanyo Special Steel Co., Ltd.
- Schmidt & Clemens GmbH & Co.
- Sheffield Steel Corp.
- Slater Steel Corp.
- St. Lawrence Steel Co.
- Stahlwerk Stahlschmidt GmbH & Co.
- Stahlwerke R. & H. Plate
- Steelmark-Eagle & Globe
- Stulz Sickles Steel Co.
- Surahammars Bruks AB
- Teledyne Pittsburgh Tool Steel
- Teledyne Vasco
- Thomson Industries Inc.
- Thyssen Edelstahlwerke Inc.
- Timken Co.
- U.N. Alloy Steel Corp.
- Uddeholm Corp.
- Westig
- Zapp, Robert, Werkstofftechnik GmbH

Ferrous electronic materials

- Armco Inc.
- Arnold Engineering Co.
- Bohler Edelstahl GmbH
- British Driver-Harris Co.
- Carpenter Technology Corp.
- Chace Precision Metals
- Hamilton Technology, Inc.
- Harrison Alloys Inc.

Ferrous electronic materials (cont'd)

- Hoskins Mfg. Co.
- Inco Alloys International, Inc
- Kanthal A.B.
- Krupp VDM
- Magnequench
- Magnetic Metals Co.
- Magnetic Shield Corp.
- Newport Steel Corp.
- Nippon Steel U.S.A. Inc.
- NKK Corp.
- Osram Sylvania Inc.
- PM Hochtemperatur-Metall GmbH
- Powdertech Corp.
- Precision Rolled Products, Inc.
- Remystahl
- Sanyo Special Steel Co., Ltd.
- Scientific Alloys Inc.
- Sheffield Forgemasters Ltd.
- Sidmar N.V.
- Stackpole Magnet Division
- Surahammars Bruks AB
- Tata Iron & Steel Co., Ltd.
- Telcon Metals Ltd.
- Thomas & Skinner Inc.
- Ulbrich Stainless Steels & Special Metals, Inc.
- Vacuumschmelze GmbH

Ferrous welding, hard facing electrodes

- Allegheny Ludlum Corp.
- Anval, Inc.
- Arcos Alloys
- Becker Bros. Carbon Corp.
- British Driver-Harris Co.
- British Steel plc
- Champion Rivet Co.
- ESAB Group
- Eutectic Corp.
- Harris, J. W., Co., Inc.
- Hobart Bros. Co.
- Inco Alloys International, Inc
- Lincoln Electric
- McKay Welding Products
- MG Industries
- Osram Sylvania Inc.
- Rankin Mfg. Co.
- Stoody Co.
- Stulz Sickles Steel Co.
- The Esab Group Inc.
- Thyssen Edelstahlwerke Inc.
- Wall Colmonoy Corp.

Aluminum and alloys

- Advanced Composite Materials Corp.
- Alcan Alluminio SpA
- Alcan Aluminium Ltd.
- Alcan Aluminum Corp.
- Alcan Wire Ltd.
- ALCOA (Aluminum Company of America)
- Alcoa/Messena
- Aluminium Francais
- Aluminium Walzwerke Singen GmbH
- Aluminum Company of Canada, Ltd.
- Aluminum Smelting & Refining Co., Inc.
- Alumix S.P.A.
- Alusuisse-Lonza Group
- American Smelting & Refining Co.
- Ampal, Inc.
- Atlantic Casting & Engineering Corp.
- Atomergic Chemetals Corp.
- BALCO
- Belmont Metals, Inc.
- Billiton International Metals BV
- Bohler Edelstahl GmbH
- Bristol Aluminum Co. (All-Lite Co.)
- British Alcan Aluminium plc
- Busch-Jager Ludenscheider Metallwerke
- Chemalloy Co., Inc.
- China Steel Corp.
- Comalco Ltd.
- Daido Steel Co. Ltd.
- Dural Aluminum Composites Corp.
- DWA Composite Specialties, Inc.
- Erbsloh, J. & A., GmbH & Co.
- ERG Materials & Aerospace Corp.
- Eutectic Corp.
- Fansteel Wellman Dynamics
- Federal-Mogul Corp.
- Fuchs, Otto Metallwerke
- Fusion Inc.
- General Motors Corp.
- Harris, J. W., Co., Inc.
- Hexcel
- Hobart Bros. Co.
- IMI plc
- Inco Alloys International, Inc
- Kaiser Aluminum & Chemical Corp.
- KM-kabelmetal AG
- Lanxide Corp.
- Lavin, R., & Sons, Inc.
- Lincoln Electric
- Linderme Tube Co.
- Mahle GmbH
- Metco Perkin-Elmer
- MG Industries
- Micron Metals Inc.
- Mueller Brass Co.
- NGK Metals Corp.
- Oster, A.J., Foils, Inc.
- Otis Elevator Co.
- Pechiney
- Phifer Wire Products
- PM Hochtemperatur-Metall GmbH
- Powder Alloy Corp.
- Precision Tube Co., Inc.
- Reading Alloys, Inc.
- Reynolds Metals Co.
- Rypac Aluminum Recycling Ltd.
- Secon Metals Corp.
- Shieldalloy Metallurgical Corp.
- Sifbronze
- Southwire Co.
- Spectro Alloys Corp.
- Sterling International Technology Ltd.
- Sulzer Plasma Technik, Inc.
- Superform USA Inc.
- Superior Die Casting Corp.

Aluminum and alloys (cont'd)

- Techni-Cast Corp.
- U.S. Reduction Co. (USCO)
- Uniform Tubes Inc.
- Unna Messingwerk, A.G.
- Valimet, Inc.
- VAW Vereinigte Aluminium-Werke AG
- Wabash Alloys, Inc.
- Wellman Dynamics Corp.

Beryllium and alloys

- Aluminum Smelting & Refining Co., Inc.
- Atomergic Chemetals Corp.
- Brush Wellman
- Williams Advanced Materials, Inc.

Bismuth and alloys

- Hoyt Alloys
- Indium Corporation of America
- MCP Systems, Inc.

Cadmium, chromium

- American Smelting & Refining Co.
- Atomergic Chemetals Corp.
- Billiton International Metals BV
- Chemalloy Co., Inc.
- Chromium Corp. of America
- Cominco Ltd.
- Engelhard Corp.
- Gecamines La General De Carrieres De Mines
- Handy & Harman
- MCP Systems, Inc.
- Outokumpu
- Reading Alloys, Inc.

Cobalt, alloys, superalloys

- Alloy Cast Products, Inc.
- Atomergic Chemetals Corp.
- Aubert et Duval
- Bohler Edelstahl GmbH
- Cannon-Muskegon Corp.
- Carboloy Inc.
- Chemalloy Co., Inc.
- Climax Specialty Metals
- Deloro Stellite Inc.
- Dewramet Ltd.
- Gecamines La General De Carrieres De Mines
- Greenville Metals, Inc.
- Haynes International, Inc.
- Howmet Corp.
- Kennametal Inc.
- MCP Systems, Inc.
- Nuclear Metals, Inc. (NMI)
- PM Hochtemperatur-Metall GmbH
- Powder Alloy Corp.
- Powdertech Corp.
- Precision Rolled Products, Inc.
- Protective Metal Alloys, Inc.
- Rankin Mfg. Co.
- Sanyo Special Steel Co., Ltd.
- Sherritt Gordon Mines Ltd.
- SPS Technologies
- Sulzer Plasma Technik, Inc.
- Telcon Metals Ltd.
- Ulbrich Stainless Steels & Special Metals, Inc.
- Vacuumschmelze GmbH
- Wall Colmonoy Corp.
- Xaloy Inc.

Copper and alloys

- A & A Brazing Alloys Inc.
- Accurate Forging Corp.
- Admiral Metals
- Alcan Alluminio SpA
- Alpha Metals
- Aluminum Smelting & Refining Co., Inc.
- American Chain & Cable
- American Crucible Products
- American Smelting & Refining Co.
- Ampco Metal
- Associated Spring Co.
- Atlantic Casting & Engineering Corp.
- Atlas Brass Foundry
- Atomergic Chemetals Corp.
- Aurora Industries
- Barker & Allen Ltd.
- Belmont Metals, Inc.
- Birmingham Battery & Metal Co. Ltd.
- Bohler Edelstahl GmbH
- British Driver-Harris Co.
- Brontel Bearing Bronze Co.
- Bunting Bearings Corp.
- Burndy Corp.
- Camden Wire Co. Inc.
- Cercast Group
- Cerro Copper Products Co.
- Cerro Metal Products Co.
- Chase Brass
- Climax Specialty Metals
- CMW Inc.
- Colonial Metals Co.
- Copper Range Co.
- Crane, Torry S., Co.
- Criterion Metals Inc.
- Delta (Manganese Bronze) Ltd.
- Delta Enfield Metals Ltd.
- Delta Extruded Metals Co. Ltd.
- Deutsche Nickel AG
- DWA Composite Specialties, Inc.
- Dyn-Metal Ltd.
- Engelhard Corp.
- Europa Metalli-LMI SpA
- Eutectic Corp.
- Federal-Mogul Corp.
- Fuchs, Otto Metallwerke
- Fusion Inc.
- G.A. Avril Co.
- General Metals Powder Co.
- Glacier Vandervell Ltd.
- Glyco-Metall-Werke
- Hackett Brass Foundry
- Handy & Harman
- Harris, Arthur & Co.
- Harris, J. W., Co., Inc.
- Harrison Alloys Inc.
- Haynes International, Inc.
- Heckford, Arthur E., Ltd.
- Hoskins Mfg. Co.
- Hussey Copper Ltd.
- IMI plc
- Inco Alloys International, Inc
- Isabellenhutte
- Jelliff Corp.
- Johnson Matthey plc
- Kanthal A.B.
- Kaydon Ring & Seal Inc.
- Kelly Foundry & Machine Co.
- Kennametal Inc.
- KM-kabelmetal AG
- Krupp VDM
- Lavin, R., & Sons, Inc.
- Leonische Drahtwerke AG
- Lincoln Electric
- Linderme Tube Co.

Copper and alloys (cont'd)

- Little Falls Alloys, Inc.
- Lunkenheimer Co.
- Magnacast Corp.
- Magnolia Metal Corp.
- Metco Perkin-Elmer
- MG Industries
- Molecu-Wire Corp.
- Mueller Brass Co.
- NGK Metals Corp.
- Nippert Co.
- Oaks Foundry Inc.
- Olin Brass
- Olin Brass, Indianapolis
- Osram Sylvania Inc.
- Oster, A.J., Foils, Inc.
- Outokumpu
- Outokumpu American Brass
- Pan-Metallgesellschaft Baumgartner GmbH
- Pechiney
- Phelps Dodge Corp.
- Philex Mining Corp.
- PM Hochtemperatur-Metall GmbH
- Powder Alloy Corp.
- Powdertech Corp.
- Precision Tube Co., Inc.
- PresMet Corp.
- Protective Metal Alloys, Inc.
- Prym-Werke, William, GmbH & Co. KG
- Pyron Corp.
- Renewal Service Inc.
- Revere Copper Products, Inc.
- Rypac Aluminum Recycling Ltd.
- Saginaw Bearing Co.
- Sandusky International
- SCM Metal Products Inc.
- Sifbronze
- Sulzer Plasma Technik, Inc.
- Techni-Cast Corp.
- Telcon Metals Ltd.
- Teledyne Rodney Metals
- Telford Shell Mouldings Ltd.
- The Miller Co.
- Tungum Hydraulics Ltd.
- Uniform Tubes Inc.
- United States Bronze Powders Inc.
- UTP. Welding Materials, Inc.
- Vacuumschmelze GmbH
- Vereinigte Metallwerke Ranshofen-Berndor
- Wall Colmonoy Corp.
- Waterbury Rolling Mills Inc.
- Waukesha Foundry, Inc.
- Weldwire Co., Inc.
- WESGO
- Western Reserve Mfg. Co.
- Wieland-Werke AG Metallwerke
- Wilkinson Co.
- Williams Advanced Materials, Inc.
- Wolverine Tube Inc.

Indium

- Aluminum Smelting & Refining Co., Inc.
- American Smelting & Refining Co.
- Atomergic Chemetals Corp.
- Indium Corporation of America
- MCP Systems, Inc.

Lead and alloys

- A & A Brazing Alloys Inc.
- Alpha Metals
- Aluminum Smelting & Refining Co., Inc.
- American Crucible Products
- American Smelting & Refining Co.
- Atomergic Chemetals Corp.
- Billiton International Metals BV
- Cadman, A.W., Manufacturing Co.
- Cerro Metal Products Co.
- Cominco Ltd.
- Division Lead Limited Partnership
- Federal-Mogul Corp.
- Fusion Inc.
- G.A. Avril Co.
- Glyco-Metall-Werke
- Hetzel & Co. Metallhuttenwerk GmbH
- Hoyt Alloys
- Indium Corporation of America
- Johnson Matthey plc
- Johnson Mfg. Co.
- Magnolia Metal Corp.
- MCP Systems, Inc.
- Metallwerk Goslar GmbH & Co. KG
- Multicore
- PresMet Corp.
- Rohr & Stolberg GmbH
- Secon Metals Corp.
- Semi-Alloy, Inc.
- Soundcoat

Lithium

- Cyprus Foote Mineral Co.

Magnesium and alloys

- Atomergic Chemetals Corp.
- Chemalloy Co., Inc.
- Dow Chemical Co.
- DWA Composite Specialties, Inc.
- ESCO Corp.
- Fansteel Wellman Dynamics
- Fuchs, Otto Metallwerke
- Greenville Metals, Inc.
- Hydro Magnesium
- Magnesium Elektron Ltd.
- Northwest Alloys, Inc.
- Pechiney
- Sanyo Special Steel Co., Ltd.
- Shieldalloy Metallurgical Corp.
- Sifbronze
- Sterling International Technology Ltd.
- Wall Colmonoy Corp.

Manganese

- Atomergic Chemetals Corp.
- Chace Precision Metals
- General Motors Corp.
- Harris, J. W., Co., Inc.
- Krupp VDM
- MCP Systems, Inc.
- Powder Alloy Corp.
- Shieldalloy Metallurgical Corp.
- Weatherly Casting & Machine Co.

Nickel, alloys, superalloys

- Allegheny Ludlum Corp.
- Ancast, Inc.
- Anor, Acieries et Forges
- Arcos Alloys
- Atomergic Chemetals Corp.
- Aubert et Duval
- Belmont Metals, Inc.
- Bohler Edelstahl GmbH
- British Driver-Harris Co.
- British Oxygen Co., Ltd.
- Cannon-Muskegon Corp.
- Carboloy Inc.
- Carlson, G.O., Inc.
- Carpenter Technology Corp.
- Certified Alloy Products, Inc.
- Chemalloy Co., Inc.
- Crane, Torry S., Co.
- Criterion Metals Inc.
- Deloro Stellite Inc.
- Deutsche Nickel AG
- Dewramet Ltd.
- Dupont, E.I., de Nemours & Co.
- Empire Steel Castings Inc.
- Enpar Sonderwerkstoffe GmbH
- Eutectic Corp.
- Fansteel VR/Wesson
- Firth-Vickers Special Steels Ltd.
- Fusion Inc.
- G & S Titanium, Inc.
- Greenville Metals, Inc.
- Harrison Alloys Inc.
- Haynes International, Inc.
- Heckford, Arthur E., Ltd.
- Hobart/TAFA Inc.
- Hoskins Mfg. Co.
- Howmet Corp.
- Inco Alloys International, Inc
- International Nickel Inc.
- Isabellenhutte
- Jelliff Corp.
- Junker, Otto, GmbH
- Kanthal A.B.
- Kennametal Inc.
- Krupp VDM
- Lincoln Electric
- Lunkenheimer Co.
- Magnetic Metals Co.
- Magnetic Shield Corp.
- Materials Development Corp.
- McGean-Rohco, Inc.
- MCP Systems, Inc.
- Metco Perkin-Elmer
- MG Industries
- Molecu-Wire Corp.
- NGK Metals Corp.
- Nuclear Metals, Inc. (NMI)
- Outokumpu
- Plymouth Tube Co., Extruded Products Div.
- PM Hochtemperatur-Metall GmbH
- Powder Alloy Corp.
- Precision Rolled Products, Inc.
- Protective Metal Alloys, Inc.
- Remystahl
- Sanyo Special Steel Co., Ltd.
- Scientific Alloys Inc.
- Shellcast Corp.
- Sherritt Gordon Mines Ltd.
- Shieldalloy Metallurgical Corp.
- Slater Steel Corp.
- Special Metals Corp.
- SPS Technologies
- Stainless Foundry & Engineering
- Stoody Co.

Nickel, alloys, superalloys (cont'd)

- Sulzer Plasma Technik, Inc.
- Technetics Corp.
- Techni-Cast Corp.
- Telcon Metals Ltd.
- Teledyne Allvac
- Teledyne Rodney Metals
- TIME GmbH
- Tri-Clover
- Ulbrich Stainless Steels & Special Metals, Inc.
- UTP. Welding Materials, Inc.
- Vacuumschmelze GmbH
- Vallourec S.A.
- Wall Colmonoy Corp.
- Waukesha Foundry, Inc.
- Weardale Steel (Wolsingham) Ltd.
- Wilkinson Co.

Nuclear, heavy metals, thorium

- Atomergic Chemetals Corp.
- Pechiney
- Pechiney Electrometallurgie
- Quantum Chemical Corp.

Precious metals

- Atomergic Chemetals Corp.
- Billiton International Metals BV
- CMW Inc.
- Cominco Ltd.
- Degussa AG
- Engelhard Corp.
- Eutectic Corp.
- Fusion Inc.
- Gecamines La General De Carrieres De Mines
- Handy & Harman
- Harris, J. W., Co., Inc.
- Indium Corporation of America
- Jelenko, J. F., & Co.

Precious metals (cont'd)

- Johnson Matthey plc
- MG Industries
- Ney, The J. M., Co.
- Osram Sylvania Inc.
- Philex Mining Corp.
- Powder Alloy Corp.
- Reactor Experiments, Inc.
- Secon Metals Corp.
- Sifbronze
- Sigmund Cohn Corp.
- Uniform Tubes Inc.
- WESGO
- Wilkinson Co.
- Williams Advanced Materials, Inc.

Refractory metals

- Atomergic Chemetals Corp.
- Billiton International Metals BV
- Carboloy Inc.
- Certified Alloy Products, Inc.
- Climax Specialty Metals
- CMW Inc.
- Dupont, E.I., de Nemours & Co.
- Engelhard Corp.
- Greenville Metals, Inc.
- H.C. Starck
- Haynes International, Inc.
- Hertel Cutting Technologies
- Hoskins Mfg. Co.
- IMI plc
- Jelenko, J. F., & Co.
- Johnson Matthey plc
- Kaydon Ring & Seal Inc.
- Kennametal Inc.
- Krupp VDM
- Kulite Tungsten Corp.
- MCP Systems, Inc.
- Metco Perkin-Elmer
- MG Industries
- Miller Thermal Technologies, Inc.
- Nuclear Metals, Inc. (NMI)
- Osram Sylvania Inc.
- Philips Elmet
- PM Hochtemperatur-Metall GmbH
- Powder Alloy Corp.
- Reading Alloys, Inc.
- Schwarzkopf Technologies
- Shieldalloy Metallurgical Corp.
- Special Metals Corp.
- Stahlwerke R. & H. Plate
- Stainless Foundry & Engineering
- Sulzer Plasma Technik, Inc.
- Teledyne Cutting Tool
- Teledyne Powder Alloys
- Teledyne Wah Chang Albany
- Uniform Tubes Inc.

Refractory metals (cont'd)

- Union Carbide Corp.
- Vereinigte Chemische Fabriken
- Wall Colmonoy Corp.
- Weardale Steel (Wolsingham) Ltd.
- Westinghouse Electric Corp.

Tin and alloys

- A & A Brazing Alloys Inc.
- Alpha Metals
- Aluminum Smelting & Refining Co., Inc.
- American Smelting & Refining Co.
- Atomergic Chemetals Corp.
- Bearium Metals Corp.
- Billiton International Metals BV
- Cadman, A.W., Manufacturing Co.
- Central Engineering & Supply Co.
- Federal-Mogul Corp.
- Fusion Inc.
- G.A. Avril Co.
- Glyco-Metall-Werke
- Hobart/TAFA Inc.
- Hoyt Alloys
- Indium Corporation of America
- Magnolia Metal Corp.
- MCP Systems, Inc.
- Metco Perkin-Elmer
- Multicore
- Telford Shell Mouldings Ltd.
- United American Metals Corp.

Titanium and alloys

- Astro Metallurgical
- Atomergic Chemetals Corp.
- Axel Johnson Metals Inc.
- Bohler Edelstahl GmbH
- CEZUS
- Chemalloy Co., Inc.
- Dupont, E.I., de Nemours & Co.
- Enpar Sonderwerkstoffe GmbH
- Fuchs, Otto Metallwerke
- G & S Titanium, Inc.
- Goldman Titanium Co.
- Hamilton Technology, Inc.
- Harvey Titanium Ltd.
- Howmet Corp.
- IMI plc
- IMI Titanium Ltd.
- Krupp VDM
- Micron Metals Inc.
- Nuclear Metals, Inc. (NMI)
- Oregon Metallurgical Corp.
- Pechiney
- Plymouth Tube Co., Extruded Products Div.
- PM Hochtemperatur-Metall GmbH
- Powder Alloy Corp.
- Precision Rolled Products, Inc.
- President Titanium & Steel Co., Inc.
- RMI Co.
- Sandvik Steel Co.
- Shieldalloy Metallurgical Corp.
- Sulzer Plasma Technik, Inc.
- Teledyne Allvac
- Teledyne Rodney Metals
- Teledyne Wah Chang Albany
- Titanium Metals Corp.
- Ulbrich Stainless Steels & Special Metals, Inc.
- Uniform Tubes Inc.
- Union Carbide Corp.
- Vallourec S.A.

Vanadium

- Atomergic Chemetals Corp.
- Reading Alloys, Inc.
- Shieldalloy Metallurgical Corp.
- Teledyne Wah Chang Albany

Zinc and alloys

- Aluminum Smelting & Refining Co., Inc.
- American Chain & Cable
- American Smelting & Refining Co.
- Atomergic Chemetals Corp.
- Billiton International Metals BV
- Cerro Metal Products Co.
- Cominco Ltd.
- Federal-Mogul Corp.
- Fusion Inc.
- Gecamines La General De Carrieres De Mines
- Handy & Harman
- Harris, J. W., Co., Inc.
- Hobart/TAFA Inc.
- Johnson Mfg. Co.
- MCP Systems, Inc.
- Metco Perkin-Elmer
- Mueller Brass Co.
- Outokumpu
- Prym-Werke, William, GmbH & Co. KG
- Revere Copper Products, Inc.
- Rohr & Stolberg GmbH
- Superior Die Casting Corp.
- Tungum Hydraulics Ltd.

Zirconium

- Atomergic Chemetals Corp.
- CEZUS
- Enpar Sonderwerkstoffe GmbH
- Goldman Titanium Co.
- Nuclear Metals, Inc. (NMI)
- Oregon Metallurgical Corp.
- Reading Alloys, Inc.
- Shieldalloy Metallurgical Corp.
- Teledyne Wah Chang Albany
- Uniform Tubes Inc.
- Westinghouse Electric Corp.

Acetals

- BASF Corp.
- DSM Engineering Plastics
- Dupont, E.I., de Nemours & Co.
- Hoechst Celanese International
- LNP Engineering Plastics
- Plastic Materials Co. Inc.
- RTP Co.
- The Polymer Corp.
- Thermofil Inc.
- TSE Industries Inc.
- Union Carbide Corp.

Acrylics

- Aristech Chemical Corp.
- BASF Corp.
- Calsak Corp.
- Cook Composites and Polymers
- Degussa Corp.
- Geon Co.
- Morton International
- Plaskolite, Inc.
- Polymer Extruded Products
- Rohm and Haas Co.
- Spartech-Resin Exchange
- Textron Specialty Materials
- TSE Industries Inc.
- Union Carbide Corp.

Cellulosics

- Scharr Industries Inc.
- Union Carbide Corp.

Epoxies

- Airtech International, Inc.
- Amalga Composites, Inc.
- American Cyanamid Co.
- Budd Co.
- Ciba Geigy Corp.
- Conap, Inc.
- Cosmic Plastics, Inc.
- Cotronics Corp.
- Epic Resins
- Ferro Corp.
- Fiber Materials Inc.
- Formulated Resins Inc.
- Franklin Fibre-Lamitex Corp.
- Glastic Corp.
- Hercules Inc.
- Hexcel
- ICI Fiberite Corp.
- Insulation Materials Inc.
- Iten Industries
- M. C. Gill Corp.
- Newport Adhesives and Composites Inc.
- Norplex Oak Inc
- NVF Co.
- Polyclad
- Polyply, Inc.
- Premix Inc.
- Products Research & Chemical Corp.
- RBC Industries Inc.
- Reichhold Chemicals Inc.
- Rogers Corp.
- Spaulding Composites Co.
- Textron Specialty Materials
- Thermoset Plastics Inc.
- 3M Co.
- TSE Industries Inc.
- Union Carbide Corp.

Polyolefins

- Allied Resinous Products, Inc.
- Arco Chemical Co.
- BASF Corp.
- Borden Packaging
- Chevron Chemical Co.
- Comalloy International Corp.
- Courtaulds Structural Composites
- Cryovac
- Dow Corning Corp.
- DSM Engineering Plastics
- Eastman Chemical Co.
- Ferro Corp.
- Fina Oil & Chemical Co.
- Himont USA Inc.
- Hoechst Celanese International
- Huntsman Chemical Corp.
- Kaneka America Corp.
- LNP Engineering Plastics

Polyolefins (cont'd)

- Monmouth Plastics, Inc.
- Neste Polymers, Inc.
- Novacor Chemicals Ltd.
- Plastic Materials Co. Inc.
- Polifil Inc.
- Polycom Huntsman Inc.
- Polymer Composites Inc.
- PPG Industries Inc.
- REXENE
- RTP Co.
- Saudi Basic Industries Corp.
- Scharr Industries Inc.
- Shell Chemical Co.
- Thermofil Inc.
- TSE Industries Inc.
- Union Carbide Corp.
- Washington Penn Plastic Co., Inc.
- Westlake Plastics Co.

Fluoropolymers, chloropolymers

- Dupont, E.I., de Nemours & Co.
- Hoechst Celanese International
- LNP Engineering Plastics
- Norplex Oak Inc.
- RTP Co.
- The Polymer Corp.

Phenolics

- American Cyanamid Co.
- Franklin Fibre-Lamitex Corp.
- Hexcel
- ICI Fiberite Corp.
- Iten Industries
- M. C. Gill Corp.
- Newport Adhesives and Composites Inc.
- Norplex Oak Inc.
- Plastics Engineering Co.
- Raymark Friction Co.

Phenolics (cont'd)

- Resinoid Engineering Corp.
- Rogers Corp.
- Spaulding Composites Co.
- TSE Industries Inc.
- Union Carbide Corp.

Polyallomers

- Aristech Chemical Corp.
- Polygon Co.

Polyamides

- AlliedSignal Inc.
- Amoco Performance Products, Inc.
- Ashley Polymers Inc.
- BASF Corp.
- Comalloy International Corp.
- Courtaulds Aerospace Inc.
- Custom Resins
- DSM Engineering Plastics
- Dupont, E.I., de Nemours & Co.
- EMS-American Grilon, Inc.
- Enichem Tecnoresine SpA
- Hoechst Celanese International
- Huls America Inc.
- Hyde, A. L., Co.
- LNP Engineering Plastics
- Miles Inc.
- Monsanto Plastics
- MRC Polymers, Inc.
- Plastic Materials Co. Inc.
- Polymer Composites Inc.
- Polymer Service Corp.
- RTP Co.
- Scharr Industries Inc.
- Spartech-Resin Exchange
- The Polymer Corp.
- Thermofil Inc.
- TSE Industries Inc.
- Ube Industries Ltd
- Wellman Inc.

Polyamides-imides

- Amoco Performance Products, Inc.
- The Polymer Corp.

Polycarbonates

- Comalloy International Corp.
- Dow Chemical Co.
- Enichem Tecnoresine SpA
- GE Plastics
- LNP Engineering Plastics
- Miles Inc.
- MRC Polymers, Inc.
- Plastic Materials Co. Inc.
- RTP Co.
- Thermofil Inc.
- TSE Industries Inc.
- Westlake Plastics Co.

Polyesters, thermoplastic

- AlliedSignal Inc.
- Amoco Performance Products, Inc.
- Bamberger Polymers, Inc.
- BASF Corp.
- Comalloy International Corp.
- Courtaulds Aerospace Inc.
- Custom Coating and Laminating Corp.
- Dupont, E.I., de Nemours & Co.
- Eastman Chemical Co.
- GAF Corp.
- GE Plastics
- Hoechst Celanese International
- LNP Engineering Plastics
- Lustro Plastics Co.
- Miles Inc.
- Polymer Composites Inc.
- PPG Industries Inc.
- Reichhold Chemicals Inc.
- RTP Co.
- Spectape Inc.
- The Polymer Corp.
- Thermofil Inc.
- TSE Industries Inc.
- Union Carbide Corp.

Polyesters, thermoset

- Airtech International, Inc.
- American Cyanamid Co.
- Applied Composites
- BASF Corp.
- BP Chemical
- Budd Co.
- Bulk Molding Compounds Inc.
- Ciba Geigy Corp.
- Cook Composites and Polymers
- Cosmic Plastics, Inc.
- Cytec Industries
- Ferro Corp.
- Franklin Fibre-Lamitex Corp.
- Glastic Corp.
- Haysite Reinforced Plastics
- Hercules Inc.
- ICI Fiberite Corp.
- Industrial Dielectrics Inc.
- Iten Industries
- Jet Moulding Compounds Ltd.
- Lustro Plastics Co.
- M. C. Gill Corp.
- MRC Polymers, Inc.
- Norplex Oak Inc
- Plastics Engineering Co.
- Polyply, Inc.
- Premix Inc.
- Reichhold Chemicals Inc.
- Rogers Corp.
- Rostone
- Scharr Industries Inc.
- Shell Chemical Co.
- Sierracin/Intrex
- Spaulding Composites Co.
- Spectape Inc.
- The Polymer Corp.

Polyketones

- Amoco Performance Products, Inc.
- BASF Corp.
- Greene, Tweed Engineered Plastics
- ICI Fiberite Corp.
- International Polymer Corp.
- LNP Engineering Plastics
- RTP Co.
- Thermofil Inc.
- TSE Industries Inc.

Polyimides

- Airtech International, Inc.
- American Cyanamid Co.
- Elf Atochem North America Inc.
- Ferro Corp.
- Fiber Materials Inc.
- Hexcel
- ICI Fiberite Corp.
- IMI-TECH Corp.
- Lenzing AG
- National Starch and Chemical Corp.
- Norplex Oak Inc.
- NVF Co.
- Polyclad
- Rogers Corp.
- Shell Chemical Co.

Polyisoimides

- Furon
- GE Plastics
- National Starch and Chemical Corp.
- RTP Co.
- Thermofil Inc.

Polyphenylenes

- BASF Corp.
- Borg-Warner Chemicals, Inc.
- GE Plastics
- Hoechst Celanese International
- Hyde, A. L., Co.
- International Polymer Corp.
- LNP Engineering Plastics
- Polymer Composites Inc.
- RTP Co.
- Thermofil Inc.

Polyurethanes

- BASF Corp.
- Ciba Geigy Corp.
- Conap, Inc.
- Deerfield Urethane, Inc.
- Dow Chemical Co.
- Dow Corning Corp.
- Formulated Resins Inc.
- Gallagher Corp.
- General Plastics Manufacturing Co.
- Geon Co.
- Hoechst Celanese International
- Lord Corp.
- Miles Inc.
- Morton International
- Polymer Composites Inc.
- Products Research & Chemical Corp.
- Quinn, K.J., & Co., Inc.
- Renosol Corp.
- RTP Co.
- Textron Specialty Materials
- Thermofil Inc.
- Thermoset Plastics Inc.
- TSE Industries Inc.
- Uniroyal Chemical Co., Inc.

Polystyrenes

- Allied Resinous Products, Inc.
- Arco Chemical Co.
- BASF Corp.
- Borg-Warner Chemicals, Inc.
- Chevron Chemical Co.
- Comalloy International Corp.
- Custom Coating and Laminating Corp.
- CYRO Industries
- Dart Polymers, Inc.
- Diamond Polymers Inc.
- Dow Chemical Co.
- Dow Corning Corp.
- DSM Engineering Plastics
- Enichem Tecnoresine SpA
- Fina Oil & Chemical Co.
- Firestone Synthetic Rubber & Latex
- GE Plastics
- GenCorp. Polymer Products
- Hoechst Celanese International
- Hsimex Southland Polymers
- Huntsman Chemical Corp.
- LNP Engineering Plastics
- Miles Inc.
- Monmouth Plastics, Inc.
- Monsanto Plastics
- Norplex Oak Inc
- Novacor Chemicals Ltd.
- Plastic Materials Co. Inc.
- Plastic Suppliers, Inc.
- RTP Co.

Polystyrenes (cont'd)

- Spartech-Resin Exchange
- The Polymer Corp.
- Thermofil Inc.
- TSE Industries Inc.
- Washington Penn Plastic Co., Inc.
- Westlake Plastics Co.

Polysulfones

- Amoco Performance Products, Inc.
- BASF Corp.
- LNP Engineering Plastics
- RTP Co.
- Thermofil Inc.
- Westlake Plastics Co.

Polyvinyl chlorides

- BASF Corp.
- Calsak Corp.
- Chris Craft Industrial Products, Inc.
- Colorite Plastics
- GenCorp. Polymer Products
- Geon Co.
- Georgia-Gulf Corp.
- Occidental Chemical Corp.
- Spectape Inc.
- Technical Plastic Extruders Inc.
- Thermofil Inc.
- Union Carbide Corp.
- Viskase Corp.

Silicones

- Dow Corning Corp.
- ICI Fiberite Corp.
- McGhan NuSil Corp.
- Novacor Chemicals Ltd.
- Polymer Technologies
- Thermoset Plastics Inc.

Ureas

- Plastics Manufacturing Co.

Thermoplastic elastomers

- Airtech International, Inc.
- BASF Corp.
- Ciba Geigy Corp.
- Concept Polymer Technologies, Inc.
- Geon Co.
- Hoechst Celanese International
- McGhan NuSil Corp.
- MCP Systems, Inc.
- Miles Inc.
- Polaroid Corp.
- RBC Industries Inc.
- RTP Co.
- Tetra Plastics Inc.
- Union Carbide Corp.
- Uniroyal Chemical Co., Inc.

Ceramics

- Cyprus Foote Mineral Co.
- P.Q. Corp.

Mechanical ceramics

- Ceradyne, Inc.

Electrical ceramics

- ALCOA (Aluminum Company of America)
- AlSiMag Technical Ceramics, Inc.
- Carborundum Co.
- Ceradyne, Inc.
- Coors Ceramics Co.
- Kanthal A.B.
- Powdertech Corp.
- Toshiba Ceramics Co., Ltd.
- Zircar Products, Inc.

Refractory ceramics

- ALCOA (Aluminum Company of America)
- Carborundum Co.
- China Metallurgical Import & Export Corp.
- Coors Ceramics Co.
- Manville Corp.
- Powder Alloy Corp.
- Toshiba Ceramics Co., Ltd.
- Zircar Products, Inc.

Engineering ceramics

- Cogebi, Inc.
- Coors Ceramics Co.
- Greenleaf Corp.
- Pechiney
- Toshiba Ceramics Co., Ltd.
- WESGO
- High Alumina and Cogebi, Inc.
- Coors Ceramics Co.
- Greenleaf Corp.
- Pechiney
- Toshiba Ceramics Co., Ltd.

High alumina and "cermets"

- AAR, Advanced Structures
- ALCOA (Aluminum Company of America)
- Alloy Technology International, Inc.
- Billiton International Metals BV
- Carboloy Inc.
- Carborundum Co.
- Coors Ceramics Co.
- Cotronics Corp.
- Fansteel VR/Wesson
- Greenleaf Corp.
- Oregon Metallurgical Corp.
- PM Hochtemperatur-Metall GmbH
- Powder Alloy Corp.
- Sulzer Plasma Technik, Inc.
- Teledyne Cutting Tool
- Toshiba Tungaloy Co., Ltd.
- Valenite Inc.
- Viking Engineered Cast Products
- WESGO
- Xaloy Inc.

Carbides

- AAR, Advanced Structures
- Atrax Co.
- Blackalloy Co. of America, Inc.
- Bohler Edelstahl GmbH
- Carbidie
- Carborundum Co.
- Carmet Co.
- Ceradyne, Inc.
- Chemalloy Co., Inc.
- CMW Inc.
- Diamond Metal Alloys
- DoAll Co.
- ERG Materials & Aerospace Corp.
- Fansteel VR/Wesson
- Greenleaf Corp.
- Howmet Corp.
- Kennametal Inc.
- Lanxide Corp.
- Metal Carbides Inc.
- Metco Perkin-Elmer
- National Machinable Carbide, Inc.
- Newcomer Products, Inc.
- Osram Sylvania Inc.
- Pechiney
- PM Hochtemperatur-Metall GmbH
- Powder Alloy Corp.
- Rankin Mfg. Co.
- Sandvik Hard Materials Ltd.
- Sandvik Steel Co.
- Sulzer Plasma Technik, Inc.
- Teledyne Cutting Tool
- Union Carbide Corp.
- Valenite Inc.
- Weardale Steel (Wolsingham) Ltd.

Nitrides

- Carborundum Co.
- Ceradyne, Inc.
- Teledyne Cutting Tool
- Toshiba Ceramics Co., Ltd.
- Union Carbide Corp.
- Valenite Inc.
- WESGO

Glass, quartz

- Courtaulds Aerospace Inc.
- Toshiba Ceramics Co., Ltd.
- Union Carbide Corp.

Carbon, graphite, diamond

- Atomergic Chemetals Corp.
- China Metallurgical Import & Export Corp.
- Ciba Geigy Corp.
- ERG Materials & Aerospace Corp.
- Graphite Metallizing Corp.
- Great Lakes Carbon Corp.
- Stackpole Magnet Division
- Superior Graphite Co.
- Teledyne Cutting Tool
- Textron Specialty Materials
- Toshiba Ceramics Co., Ltd.
- Union Carbide Corp.
- Valenite Inc.
- Wellman, S.K., Corp.

Mica

- P.Q. Corp.

Oxides

- AlSiMag Technical Ceramics, Inc.
- Ceradyne, Inc.
- China Metallurgical Import & Export Corp.
- Cotronics Corp.
- Lanxide Corp.
- Sulzer Plasma Technik, Inc.
- Toshiba Ceramics Co., Ltd.
- Zircar Products, Inc.

Fibers

- Allied Fibers
- American Cyanamid Co.
- Ciba Geigy Corp.
- Courtaulds Aerospace Inc.
- Hercules Inc.
- Hoechst Celanese International
- ICD Group Inc.
- Owens Corning Fiberglass Corp.
- Soundcoat
- Textron Specialty Materials
- Union Carbide Corp.
- Zoltek

Fillers, additives

- ALCOA (Aluminum Company of America)
- Genstar Stone Products Co.
- Georgia Marble Co.
- GREFCO, Inc.
- NYCO
- Pechiney
- Pfizer Minerals, Pigments & Metals Div.
- PQ Corp.
- The Andersons
- Zeelan Industries Inc.

Index to Producer Listing

INDEX TO MATERIAL LISTING

For your convenience, in this index we further break down the engineering materials categories used in this directory. To locate a producer of gold, for example, you are directed to the "Precious metals" heading. To locate a producer of polypropylene, you are directed to the "Polyolefins" heading. Among the producers listed under the appropriate heading will be at least one source of the specific material you seek.

UPDATE FORM

If you have any additions or corrections to this guide, please send them to us so that we may improve future editions.

Company name: ________________________________
Contact name: ________________________________
Company address: ________________________________
Company phone: ________________________________
Company fax: ________________________________

Additional Locations
Company name: ________________________________
Contact name: ________________________________
Company address: ________________________________
Company phone: ________________________________
Company fax: ________________________________

(use additional page up to 20 locations)

Materials Produced:

- ☐ Carbon steels
- ☐ Carbon steels, cast
- ☐ Carbon steels, mill shapes
- ☐ Carbon steels, sheet, strip
- ☐ Carbon steels, plate
- ☐ Carbon steels, forged
- ☐ Carbon steels, powder
- ☐ Alloy steels
- ☐ Alloy steels, cast
- ☐ Alloy steels, mill shapes
- ☐ Alloy steels, sheet, strip
- ☐ Alloy steels, plate
- ☐ Alloy steels, forged
- ☐ Alloy steels, powder
- ☐ Stainless steels
- ☐ Stainless steels, cast
- ☐ Stainless steels, mill shapes
- ☐ Stainless steels, sheet. strip
- ☐ Stainless steels, plate
- ☐ Stainless steels, forged
- ☐ Stainless steels, powder
- ☐ Ferroalloys, pure iron
- ☐ Iron-base superalloys
- ☐ Cast iron
- ☐ Tool steels
- ☐ Ferrous electronic materials
- ☐ Ferrous welding, hard facing electrodes
- ☐ Aluminum and alloys
- ☐ Antimony
- ☐ Beryllium and alloys
- ☐ Bismuth and alloys
- ☐ Cadmium, chromium
- ☐ Cobalt, alloys, superalloys
- ☐ Copper and alloys
- ☐ Indium
- ☐ Lead and alloys
- ☐ Lithium
- ☐ Magnesium and alloys
- ☐ Manganese
- ☐ Nickel, alloys, superalloys
- ☐ Nuclear, heavy metals, thorium
- ☐ Precious metals
- ☐ Refractory metals
- ☐ Tin and alloys
- ☐ Titanium and alloys
- ☐ Vanadium
- ☐ Zinc and alloys
- ☐ Zirconium
- ☐ Acetals
- ☐ Acrylics
- ☐ Cellulosics
- ☐ Epoxies
- ☐ Polyolefins
- ☐ Fluoropolymers, chloropolymers
- ☐ Furan
- ☐ Phenolics
- ☐ Polyallomers
- ☐ Polyamides
- ☐ Polyamides-imides
- ☐ Polycarbonates
- ☐ Polyesters, thermoplastic
- ☐ Polyesters, thermoset
- ☐ Polyketones
- ☐ Polyimides
- ☐ Polyisoimides
- ☐ Polyisocyanurates
- ☐ Polymethylpentenes
- ☐ Polyphenylenes
- ☐ Polyurethanes
- ☐ Polystyrenes
- ☐ Polysulfones
- ☐ Polyvinyl chlorides
- ☐ Silicones
- ☐ Ureas
- ☐ Thermoplastic elastomers
- ☐ Ceramics
- ☐ Mechanical ceramics
- ☐ Electrical ceramics
- ☐ Refractory ceramics
- ☐ Engineering ceramics
- ☐ High alumina and "cermets"
- ☐ Carbides
- ☐ Nitrides
- ☐ Glass, quartz
- ☐ Carbon, graphite, diamond
- ☐ Mica
- ☐ Oxides
- ☐ Fibers
- ☐ Fillers, additives

(Please check any materials that apply to your company)

(Please fold along dashed lines and mail)

Return to: **ASM International**
Center for Materials Data
Materials Park, OH 44073